# FRONTIERS IN AGRICULTURAL RESEARCH

## Food, Health, Environment, and Communities

Committee on Opportunities in Agriculture

BOARD ON AGRICULTURE AND NATURAL RESOURCES
DIVISION ON EARTH AND LIFE STUDIES

NATIONAL RESEARCH COUNCIL
*OF THE NATIONAL ACADEMIES*

THE NATIONAL ACADEMIES PRESS
Washington, D.C.
**www.nap.edu**

**THE NATIONAL ACADEMIES PRESS**   500 Fifth Street, NW   Washington, DC 20001

NOTICE: The project that is the subject of this report was approved by the Governing Board of the National Research Council, whose members are drawn from the councils of the National Academy of Sciences, the National Academy of Engineering, and the Institute of Medicine. The members of the committee responsible for the report were chosen for their special competences and with regard for appropriate balance.

This study was supported by Contract/Grant No. 59-0790-9-172 between the National Academy of Sciences and the US Department of Agriculture. Any opinions, findings, conclusions, or recommendations expressed in this publication are those of the authors and do not necessarily reflect the views of the organizations or agencies that provided support for the project.

International Standard Book Number: 0-309-08494-6
Library of Congress Control Number: 2002115520

Additional copies of this report are available from National Academies Press, 500 Fifth Street, NW, Lockbox 285, Washington, DC 20055; (800) 624-6242 or (202) 334-3313 (in the Washington metropolitan area); Internet, http://www.nap.edu

Suggested citation: National Research Council, 2003. *Frontiers in Agricultural Research: Food, Health, Environment, and Communities.* Committee on Opportunities in Agriculture (Washington, DC: National Academies Press).

Copyright 2003 by the National Academy of Sciences. All rights reserved.

Printed in the United States of America

# THE NATIONAL ACADEMIES
*Advisers to the Nation on Science, Engineering, and Medicine*

The **National Academy of Sciences** is a private, nonprofit, self-perpetuating society of distinguished scholars engaged in scientific and engineering research, dedicated to the furtherance of science and technology and to their use for the general welfare. Upon the authority of the charter granted to it by the Congress in 1863, the Academy has a mandate that requires it to advise the federal government on scientific and technical matters. Dr. Bruce M. Alberts is president of the National Academy of Sciences.

The **National Academy of Engineering** was established in 1964, under the charter of the National Academy of Sciences, as a parallel organization of outstanding engineers. It is autonomous in its administration and in the selection of its members, sharing with the National Academy of Sciences the responsibility for advising the federal government. The National Academy of Engineering also sponsors engineering programs aimed at meeting national needs, encourages education and research, and recognizes the superior achievements of engineers. Dr. Wm. A. Wulf is president of the National Academy of Engineering.

The **Institute of Medicine** was established in 1970 by the National Academy of Sciences to secure the services of eminent members of appropriate professions in the examination of policy matters pertaining to the health of the public. The Institute acts under the responsibility given to the National Academy of Sciences by its congressional charter to be an adviser to the federal government and, upon its own initiative, to identify issues of medical care, research, and education. Dr. Harvey V. Fineberg is president of the Institute of Medicine.

The **National Research Council** was organized by the National Academy of Sciences in 1916 to associate the broad community of science and technology with the Academy's purposes of furthering knowledge and advising the federal government. Functioning in accordance with general policies determined by the Academy, the Council has become the principal operating agency of both the National Academy of Sciences and the National Academy of Engineering in providing services to the government, the public, and the scientific and engineering communities. The Council is administered jointly by both Academies and the Institute of Medicine. Dr. Bruce M. Alberts and Dr. Wm. A. Wulf are chair and vice chair, respectively, of the National Research Council.

**www.national-academies.org**

# COMMITTEE ON OPPORTUNITIES IN AGRICULTURE

LAURIAN J. UNNEVEHR,[1] *Chair*, University of Illinois, Urbana-Champaign, Illinois
FRANKLIN M. LOEW,[2] Becker College, Worcester, Massachusetts
RANSOM L. BALDWIN, JR., University of California, Davis, California
ROGER N. BEACHY, Donald Danforth Plant Science Center, St. Louis, Missouri
CAROLYN BRANCH BROOKS, University of Maryland, Eastern Shore, Princess Anne, Maryland
ELIZABETH A. CHORNESKY, The Nature Conservancy, Santa Cruz, California
EDWARD A. HILER, Texas A&M University, College Station, Texas
WALLACE E. HUFFMAN, Iowa State University, Ames, Iowa
LONNIE J. KING, Michigan State University, East Lansing, Michigan
LAWRENCE N. KUZMINSKI, Consultant, Duxbury, Massachusetts
WILLIAM B. LACY, University of California, Davis, California
THOMAS L. LYON, Cooperative Resources International, Shawano, Wisconsin
KRISTEN MCNUTT, Consumer Choices, Inc., Santa Cruz, California
WILLIAM L. OGREN (retired), Hilton Head Island, South Carolina
DAVID PIMENTEL, Cornell University, Ithaca, New York
ROBERT REGINATO (retired), Chandler, Arizona
JOHN W. SUTTIE, University of Wisconsin, Madison, Wisconsin

*Staff*

CLARA COHEN, *Study Director* (since September 2001)
DAVID MEEKER, *Study Director* (until September 2001)
NORMAN GROSSBLATT, *Senior Editor*
LUCYNA KURTYKA, *Program Officer* (until July 2000)
MICHAEL R. KISIELEWSKI, *Research Assistant*
HEATHER CHRISTIANSEN, *Research Associate* (until September 2001)
JOE ESPARZA, *Project Assistant*
LAURA BOSCHINI, *Project Assistant* (until September 2001)

---

[1] Chair from March 2002 to December 2002.
[2] Chair from July 2000 to March 2002.

## COMMITTEE ON OPPORTUNITIES IN AGRICULTURE SUBCOMMITTEE ON ECONOMIC AND SOCIAL DEVELOPMENT IN A GLOBAL CONTEXT

RAY GOLDBERG, *Chair*, Harvard Business School, Boston, Massachusetts
JULIAN ALSTON, University of California, Davis, California
LAWRENCE M. BUSCH, Institute for Food and Agricultural Standards, Michigan State University, East Lansing, Michigan
CHRISTINE BRUHN, University of California, Davis, California
PIERRE CROSSON, Resources for the Future, Inc., Washington, DC
BRIAN HALWEIL, WorldWatch Institute, Washington, DC
FRED HARRISON, JR., Fort Valley State University, Fort Valley, Georgia
CAROL KEISER, C-BAR Cattle Company, Inc., Champaign, Illinois
TERRY L. ROE, University of Minnesota, St. Paul, Minnesota
LAURIAN J. UNNEVEHR, University of Illinois, Urbana-Champaign, Illinois

## COMMITTEE ON OPPORTUNITIES IN AGRICULTURE SUBCOMMITTEE ON ENVIRONMENTAL QUALITY AND NATURAL RESOURCES

G. PHILIP ROBERTSON, *Chair*, Michigan State University, Hickory Corners, Michigan
JENNY BROOME, University of California, Davis, California
ELIZABETH CHORNESKY, The Nature Conservancy, Santa Cruz, California
JANE FRANKENBERGER, Purdue University, West Lafayette, Indiana
PAUL JOHNSON, Oneota Slopes Farm, Decorah, Iowa
MARK LIPSON, Organic Farming Research Foundation, Santa Cruz, California
JOHN MIRANOWSKI, Iowa State University, Ames, Iowa
JAMES MOSELEY,[1] Infinity Pork, Clark Hills, Indiana
ELIZABETH OWENS, Monsanto Company, Chesterfield, Missouri
DAVID PIMENTEL, Cornell University, Ithaca, New York
LORI ANN THRUPP, Environmental Protection Agency, San Francisco, California

---

[1] Resigned April 9, 2001.

## COMMITTEE ON OPPORTUNITIES IN AGRICULTURE
## SUBCOMMITTEE ON FOOD AND HEALTH

SUSAN HARLANDER, *Chair*, BIOrational Consultants, Inc., New Brighton, Minnesota
LESTER M. CRAWFORD, JR.,[1] Association of American Veterinary Colleges, Washington, DC
JOAN R. DAVENPORT, Washington State University, Prosser, Washington
REBECCA DOYLE,[2] Andrews, Doyle, and Associates, Gillespie, Illinois
DONALD N. DUVICK, Iowa State University, Johnston, Iowa
JOSEPH JEN,[3] California State Polytechnic University, San Luis Obispo, California
JOHN B. KANEENE, Michigan State University, East Lansing, Michigan
LAWRENCE N. KUZMINSKI, Duxbury, Massachusetts
ARNO G. MOTULSKY, University of Washington, Seattle, Washington
DAVID L. PELLETIER, Cornell University, Ithaca, New York
JEAN A. T. PENNINGTON, National Institutes of Health, Department of Health and Human Services, Bethesda, Maryland
MAX ROTHSCHILD, Iowa State University, Ames, Iowa
ANDREW SCHMITZ, University of Florida, Gainesville, Florida
JOHN W. SUTTIE, University of Wisconsin, Madison, Wisconsin

---

[1] Resigned February 28, 2002.
[2] Resigned July 20, 2001.
[3] Resigned April 9, 2001.

# BOARD ON AGRICULTURE AND NATURAL RESOURCES

HARLEY W. MOON, *Chair*, Iowa State University, Ames, Iowa
SANDRA BARTHOLMEY, Quaker Oats Company, Barrington, Illinois
DEBORAH BLUM, University of Wisconsin, Madison, Wisconsin
ROBERT B. FRIDLEY, University of California, Davis, California
BARBARA P. GLENN, Federation of Animal Science Societies, Bethesda, Maryland
LINDA F. GOLODNER, National Consumers League, Washington, DC
W. R. GOMES, University of California, Oakland, California
PERRY R. HAGENSTEIN, Institute for Forest Analysis, Planning, and Policy, Wayland, Massachusetts
CALESTOUS JUMA, Harvard University, Cambridge, Massachusetts
JANET C. KING, University of California, Davis, California
WHITNEY MACMILLAN (retired), Cargill, Inc., Minneapolis, Minnesota
TERRY L. MEDLEY, DuPont BioSolutions Enterprise, Wilmington, Delaware
ALICE N. PELL, Cornell University, Ithaca, New York
SHARRON S. QUISENBERRY, Montana State University, Bozeman, Montana
NANCY J. RACHMAN, Novigen Sciences, Inc., Washington, DC
SONYA B. SALAMON, University of Illinois, Urbana-Champaign, Illinois
G. EDWARD SCHUH, University of Minnesota, Minneapolis, Minnesota
BRIAN J. STASKAWICZ, University of California, Berkeley, California
JACK WARD THOMAS, University of Montana, Missoula, Montana
JAMES H. TUMLINSON, Agriculture Research Service, US Department of Agriculture, Gainesville, Florida
B.L. TURNER, Clark University, Worcester, Massachusetts

*Staff*

DAVID MEEKER, *Director* (until September 2001)
JULIE ANDREWS, *Administrative Assistant* (until May 2002)

# Preface

Rapid and dramatic social, economic, and technologic changes have occurred in the food and agricultural sector during the last 30 years. These include increased global competition, the advent of biotechnology and precision production, changes in intellectual property rights, increased product differentiation, greater demand for ecosystem services from agriculture, and changes in farm and market structure. Thirty years have passed since the publication of the 1972 report of the National Research Council Committee on Research Advisory to the US Department of Agriculture (USDA), and the time is now ripe for reflection on the progress and future directions of federally funded agricultural research, education, and extension. The USDA asked the Research Council's Board on Agriculture and Natural Resources (BANR) to conduct a study to examine and evaluate the quality of research conducted in USDA's Research, Education, and Economics (REE) mission area and to provide recommendations for future research. The request responded to a congressional mandate in the 1998 Agricultural Research, Extension, and Education Reform Act for the National Academy of Sciences to conduct a study of the role and mission of federally funded agricultural research, extension, and education (see Appendix A).

To respond to the request, BANR convened four ad hoc study panels—a synthesis committee and three subcommittees—addressing specific components of agricultural research, education, and extension: food and fiber supply, food safety, diet, and nutrition; environmental quality and natural resources; and economic and social development in a global context. The panels represent a wide array of expertise and include those with knowledge of public and private agricultural research and those who use or are affected by the results of the research. Many members of the panels have experience in and understanding of the historical context of publicly funded agricultural research; others are in basic-science fields and have little direct experience with this system. Thus, the panels represented diverse viewpoints on the role and relevance of research, the means of achieving future research goals, and the effects of research results.

The Committee on Opportunities in Agriculture was given the following charge:

1. Drawing in part on previous National Research Council work, the study will include collection, review, and assessment of data on agricultural research and its operating environment.
2. The historical background of agricultural research, education, and economics will be considered, and changes in US needs and priorities will be described.
3. Programmatic and functional complementarities among the four REE research agencies will be examined, and the relevance of agency research to current and proposed national priorities will be evaluated.
4. Current capacity in research, education, and extension will be assessed, and scientific strengths and gaps in federally funded agricultural research efforts will be identified.
5. Research quality will be evaluated for content, relevance, effectiveness, and outcome with regard to the Government Performance and Results Act of 1993.
6. Quality standards, the use of peer review and external advice, resource allocation (including formula funds), and collaborative and interdisciplinary research will be examined.
7. Recommendations will be provided on the future role of federally funded agricultural research, future research opportunities and directions, the setting of relevant research priorities, gaps or weaknesses in the federal agricultural research system, and the strengthening of programmatic, structural, or management components of agricultural research, extension, and education to ensure responsiveness to future national needs.

The panels focused on the changing context of agricultural research and the widening array of potential benefits to society as the starting point for their review. Thus, the committee and its subcommittees first identified the important research opportunities and then reviewed the REE agencies' operations to see how they could take greater advantage of the opportunities.

The committee established statements of task for each of the three subcommittees (see Appendix B), which focused on particular issues in their own subjects. The subcommittees generated white papers that provided input into the final report; the white papers were particularly useful for identifying cutting-edge research opportunities peculiar to the three broad subjects and for identifying ways to address the opportunities.

The four panels gathered data from various sources. A public workshop was held in May 2001, and many stakeholders and clients of REE research participated (see Appendix C). The panels requested data from REE agencies. They considered the scholarly and gray literature, including REE Web sites, the Current Research Information System, budget data, agency performance reports and strategic plans, and previous National Research Council reviews of USDA research: *Publicly Funded Agricultural Research and the Changing Structure of US Agriculture* (2002), *National Research Initiative: A Vital Competitive Grants Program in Food, Fiber, and Natural Resources Research* (2000), *Sowing the Seeds of Change: Informing Public Policy in the Economic Research Service of USDA* (1999), *Colleges of Agriculture at the Land Grant Universities: Public Service and Public Policy* (1996), *Colleges of Agriculture at the Land Grant Universities: A Profile* (1995), *Investing in the National Research Initiative: An Update of the Competitive Grants Program of the US Department of Agriculture* (1994), *Investing in Research: A Proposal to Strengthen the Agricultural, Food, and Environmental System* (1989). And they conducted telephone interviews with administrators of the four REE agencies (Appendix D), administrators of other USDA agencies (Appendix E), and staff of the Office of Human Resources, ARS Office of International Programs, and of the ARS Office of Technology Transfer. The panels are grateful to the USDA staff and administrators for sharing their time, information, and experience to assist them in understanding how the unique and complex REE enterprise functions.

Chapter 1 presents the committee's vision of the future of federally funded agricultural research and sets the stage for the rest of the report. Chapter 2 provides background on the REE mission area and agencies. Chapter 3 describes the key research frontiers for the future, which motivate the review of REE policies, organization, and processes in Chapters 4 through 7, which deal, respectively, with research strategy, collaboration, quality and impact assurance in the REE agencies, and REE capacity. Those four chapters provide recommendations for changes in resource allocation, research leadership, relevance assurance, and collaborative partnerships.

The study panels hope that Congress and REE will find the recommendations and analysis in the report useful in crafting future agricultural research policy that responds to a broader array of national needs.

Laurian Unnevehr, *Chair*
Committee on Opportunities in Agriculture

# Acknowledgments

The committee is extremely grateful to numerous people who contributed time and expertise during the development of this report.

The committee thanks especially the following, who participated in its public meeting in May 2001:

JOHN B. ADAMS, National Milk Producers Federation
DICK AMERMAN, Agricultural Research Service, US Department of Agriculture
WALTER ARMBRUSTER, Farm Foundation
JILL AUBURN, Sustainable Agriculture Research and Education Program, US Department of Agriculture
KATE CLANCY, Winrock International
NEIL COWEN, Dow Agro
MONTAGUE DEMMENT, University of California, Davis
STEVE DERRENBACHER, Northeast Pasture Research and Extension Consortium
ROBERT DONALDSON, George Washington University
JERE DOWNING, Cranberry Institute
ROBERT EARL, National Food Processors Association
CORNELIA FLORA, North Central Regional Center for Rural Development, Iowa State University
BRUCE L. GARDNER, University of Maryland
JERRY R. GILLESPIE, Joint Institute for Food Safety Research
KARL GLASENER, CoFARM
BARBARA P. GLENN, Federation of Animal Science Societies

CLARE M. HASLER, University of Illinois at Urbana-Champaign
ROBERT HEDBERG, Weed Science Society of America
RICHARD HERRETT, Agricultural Research Institute
WALTER A. HILL, Tuskegee University
MYRON JOHNSRUD, National Association of State Universities and Land Grant Colleges
MARLYN JORGENSEN, Jorg-Anna Farms Partnership
CHARLES KRUEGER, The Pennsylvania State University
RATTAN LAL, Ohio State University
GERALD LARSON, Office of Budget and Program Analysis, US Department of Agriculture
KIM LEVAL, Center for Rural Affairs, Consortium for Sustainable Agriculture Research
ESTHER MYERS, American Dietetic Association
GEORGE W. NORTON, Virginia Tech
MICHAEL O'NEILL, Cooperative State Research, Education, and Extension Service, US Department of Agriculture
FRAN PIERCE, Precision Agriculture Center, Washington State University
DONNA PORTER, Congressional Research Service
CHARLES RIEMENSCHNEIDER, Food and Agriculture Organization of the United Nations
CHARLES SCIFRES, Texas Agricultural Experiment Station, Texas A&M University
STEPHANIE SMITH, Institute for Food Technologists
CAROLINE SMITH-DEWAAL, Center for Science in the Public Interest
ANN SORENSEN, American Farmland Trust
ROGER A. SUNDE, University of Missouri at Columbia
LOUIS E. SWANSON, Colorado State University
ANNE SYDNOR, Food Marketing Institute
RANDALL TORGERSON, Rural Business Cooperative Service, US Department of Agriculture
TAMARA WAGESTER, The Council on Food, Agriculture, and Resource Economics
C. MIKE WILLIAMS, Animal and Poultry Waste Management Center, North Carolina State University
TERRY WOLF, National Coalition for Food and Agricultural Research
LAREESA WOLFENBARGER, University of Nebraska, Omaha
CATHERINE E. WOTEKI, Iowa State University

The committee also thanks the following individuals who provided written statements:

CURRY ANDERSON, Nebraska Energy Crop Association

BOB BUDD, The Nature Conservancy
THOMAS CARUSO, Virginia Tech
CHARLES CURTIN, Gray Ranch and Malpai Borderlands Group
HOWARD GARRISON, Federation of American Societies for Experimental Biology
JAN HOPMANS, University of California, Davis
ALLISON JONES, National Alliance of Independent Crop Consultants
NADINE LYMN, Ecological Society of America
COLIN KALTENBACH, University of Arizona
FRED KIRSCHENMANN, Leopold Center for Sustainable Agriculture, Iowa State University
RICHARD KNIGHT, Colorado State University
HERBERT KNUDSEN, Natural Fibers Corporation
ROBERT LUXMOORE, Oak Ridge National Laboratory
PAM MARRONE, AgraQuest, Inc.
ADELE MORRIS, US Department of the Treasury
JOHN NICHOLAIDES, Soil Science Society of America
DEANNA OSMOND, North Carolina State University
GEORGE RUYLE, University of Arizona
TIMOTHY SEASTEDT, University of Colorado, Boulder
ERNIE SHEA, National Association of Conservation Districts
DOUGLAS SLOTHOWER, American Society of Farm Managers and Rural Appraisers
CLIFFORD SNYDOR, Potash and Phosphate Institute
JANICE THIES, Cornell University
RAY WEIL, University of Maryland
SHARON WEISS, International Life Sciences Institute

The committee also thanks those who spoke at its meetings:

RAY GOLDBERG, Harvard Business School
JOSEPH JEN, Research, Education, and Economics, US Department of Agriculture
EILEEN KENNEDY, International Life Sciences Institute
VICTOR LECHTENBERG, Purdue University
SARA MAZIE, Research, Education, and Economics, US Department of Agriculture
SAMUEL SMITH, Washington State University

The committee thanks the administrators of US Department of Agriculture agencies for their candor, suggestions, and input in telephone interviews:

DAVID ACHESON, Food Safety and Inspection Service

BOBBY ACORD, Animal and Plant Health Inspection Service
NATE BAUER, Food Safety and Inspection Service
JAMES BLAYLOCK, Economic Research Service
RON BOSECKER, National Agricultural Statistics Service
GEORGE BRALEY, Food and Nutrition Service
LAWRENCE CLARK, Natural Resources Conservation Service
JOHN CLIFFORD, Animal and Plant Health Inspection Service
NEILSON CONKLIN, Economic Research Service
RON DEHAVEN, Animal and Plant Health Inspection Service
RICHARD DUNKLE, Animal and Plant Health Inspection Service
ALBERTA FROST, Food and Nutrition Service
PHIL FULTON, Economic Research Service
MARGARET GLAVIN, Food Safety and Inspection Service
FLOYD HORN, Agricultural Research Service
COLIEN HEFFERAN, Cooperative State Research, Education, and Extension Service
JAMES LITTLE, Farm Services Agency
SUSAN OFFUTT, Economic Research Service
KATHERINE SMITH, Economic Research Service
RALPH STAFKO, Food Safety and Inspection Service

The committee acknowledges those who assisted the National Research Council staff during preparation of the report by providing information and statistics to the committee:

RICHARD BRENNER, Agricultural Research Service, US Department of Agriculture
KAREN BROWNELL, Office of Human Resources, Research, Education, and Economics, US Department of Agriculture
GEORGE COOPER, Cooperative State Research, Education, and Extension Service, US Department of Agriculture
KELLY DAY-RUBENSTEIN, Economic Research Service, US Department of Agriculture
JANICE GOODWIN, National Agricultural Statistics Service, US Department of Agriculture
RALPH HEIMLICH, Economic Research Service, US Department of Agriculture
EILEEN HERRERA, Agricultural Research Service, US Department of Agriculture
VIRGINIA HOUK, US Environmental Protection Agency
KEI KOIZUMI, American Association for the Advancement of Science
MEL MATHIAS, Cooperative State Research, Education, and Extension Service, US Department of Agriculture

*Acknowledgments* xix

SARA MAZIE, Office of the Undersecretary, Research, Education, and Economics, US Department of Agriculture
ARLYNE MEYERS, Office of International Programs, Agricultural Research Service, US Department of Agriculture
MARCIA MOORE, Office of Scientific Quality Review, Agricultural Research Service, US Department of Agriculture
DAVID RUST, Agricultural Research Service, US Department of Agriculture
RICHARD SCHUCHARDT, National Agricultural Statistics Service, US Department of Agriculture
DEBORAH SHEELY, Cooperative State Research, Education, and Extension Service, US Department of Agriculture
VICKI SMITH, Economic Research Service, US Department of Agriculture
DENNIS UNGLESBEE, Cooperative State Research, Education, and Extension Service, US Department of Agriculture
RICHARD WYATT, National Institutes of Health, US Department of Health and Human Services

The committee is extremely grateful to the staff members of the National Research Council Board on Agriculture and Natural Resources for their efforts throughout the study process and in the preparation of this report. The committee acknowledges Gregory Symmes, Associate Executive Director of the National Research Council Division on Earth and Life Studies, for providing guidance in the latter stages of the project. We are especially grateful to Clara Cohen, Study Director, for her extraordinary efforts and leadership in coordinating the work of all the committees and for overseeing the preparation and completion of this report.

This report has been reviewed in draft form by individuals chosen for their diverse perspectives and technical expertise, in accordance with procedures approved by the National Research Council's Report Review Committee. The purpose of this independent review is to provide candid and critical comments that will assist the institution in making its published report as sound as possible and to ensure that the report meets institutional standards of objectivity, evidence, and responsiveness to the study charge. The review comments and draft manuscript remain confidential to protect the integrity of the deliberative process. We wish to thank the following individuals for their review of this report:

JOANNE CHORY, Salk Institute for Biological Studies
RALPH CICERONE, University of California, Irvine
RONALD ESTABROOK, University of Texas Southwestern Medical Center
ROBERT B. FRIDLEY, University of California, Davis
BETH LAUTNER, National Pork Board

ALAN I. LESHNER, American Association for the Advancement of Science
LINDA LOBAO, Ohio State University
NOREEN NOONAN, National Space Science and Technology Center
PHILIP PARDEY, University of Minnesota
BARBARA SCHNEEMAN, University of California, Davis
SAM SMITH, Washington State University
M.S. SWAMINATHAN, M.S. Swaminathan Research Foundation
MICHAEL TAYLOR, Resources for the Future
CAROL TUCKER FOREMAN, Consumer Federation of America

Although the reviewers listed above have provided many constructive comments and suggestions, they were not asked to endorse the conclusions or recommendations, nor did they see the final draft of the report before its release. The review of this report was overseen by Enriqueta Bond, Burroughs Wellcome Fund, and James Zuiches, Washington State University. Appointed by the National Research Council, they were responsible for making certain that an independent examination of this report was carried out in accordance with institutional procedures and that all review comments were carefully considered. Responsibility for the final content of this report rests entirely with the author committee and the institution.

# Contents

EXECUTIVE SUMMARY 1
   The Changing Context of Agricultural Research, 2
   A Vision of Agricultural Research, 3
   Frontiers in Agricultural Research, 4
   Setting the Research Strategy, 6
   Collaboration, 9
   Quality and Impact Assurance, 10
   REE Capacity, 11
   Looking to the Future, 14

1  VISION AND LEADERSHIP 15
   Changing Public Attitudes and Needs, 16
   Recent Innovations in Science and Technology, 19
   A Vision for the Future, 21
   Summary, 24
   References, 25

2  THE US DEPARTMENT OF AGRICULTURE RESEARCH, 27
   EDUCATION, AND ECONOMICS MISSION AREA
   Reorganization of the US Department of Agriculture, 28
   Functions and Strategic Objectives of the Research, Education, and
      Economics Mission Area, 30
   Research, Education, and Economics Agencies, 30
   Summary, 36
   References, 36

## 3 RESEARCH FRONTIERS 38
Globalization, 39
Emerging Pathogens and Other Hazards in the
   Food-Supply Chain, 42
Nutrition and Human Health, 45
Environmental Stewardship, 49
Quality of Life in Rural Communities, 52
Advancing the Frontiers, 54
Summary, 61
References, 61

## 4 SETTING THE RESEARCH STRATEGY 67
Funding Sources and Trends, 68
REE and Agency Decision-Making, 72
Summary, 92
References, 92

## 5 COLLABORATION 96
Multidisciplinary Research, 96
Collaboration within REE, 100
Collaboration in the Federal Government, 102
International Collaboration, 105
Collaboration with the Private Sector, 108
Future Strategies to Manage Public–Private Collaboration, 114
Summary, 115
References, 115

## 6 QUALITY AND IMPACT ASSURANCE IN THE REE AGENCIES 119
Quality Assurance, 119
Impact Assessment, 133
Summary, 142
References, 142

## 7 REE CAPACITY 146
Organizational Capacity, 146
Professional Skills, Expertise, and Training, 150
Information Capacity: REE Efforts in Data Management,
   Collection and Sharing, 161
Infrastructure Capacity: Research Facilities, 165
Summary, 167
References, 167

| 8 | CODA | 169 |

APPENDIXES

| A | S.1150.1998. Agricultural Research, Extension, and Education Reform Act of 1998 | 173 |
| B | Subcommittee Statements of Task | 175 |
| C | A National Research Council Public Workshop | 177 |
| D | REE Administrator Interviews | 182 |
| E | Action-Agency Administrator Interviews | 184 |
| F | Agricultural-Research Funding | 186 |
| G | REE Dissemination and Outreach Efforts | 204 |

| ABOUT THE AUTHORS | 217 |
| ABOUT THE SUBCOMMITTEES | 225 |

# Tables, Figures, and Boxes

**TABLES**

| | |
|---|---|
| 4-1 | REE Mechanisms for Ensuring Stakeholder Input, page 86 |
| 4-2 | Membership Categories, Represented in the National Agricultural Research, Extension, Education, and Economics Advisory Board, 87 |
| 5-1 | Visiting Scientists at ARS, 1998–2001, 105 |
| 5-2 | USDA Technology Transfer Activities, 1987–2000, 109 |
| 6-1 | Summary of REE Quality-Assurance Mechanisms, 121 |
| 6-2 | Results of February 2000–August 2001 Review of Six ARS National Programs, 125 |
| 6-3 | 1999–2000 Intramural (ARS and ERS) Recipients of Major Awards Sponsored by External Organizations, 132 |
| 6-4 | World Institutional Rankings in Select Fields, by Total Citations, 1991–2001, 134 |
| 6-5 | Internal Rates of Return from US Public-Sector Agricultural Research, 136 |
| 6-6 | Internal Rates of Return from US Extension, 137 |
| 7-1 | REE Professional Employment in Science-Related Occupations, as of June 10, 2001, 151 |
| 7-2 | Demographic Composition of REE Technical Staff, 152 |
| 7-3 | ARS Postdoctoral Employment, 157 |
| 7-4 | Funding Levels for the ARS Postdoctoral Program, 1985–2002, 157 |
| 7-5 | Summary of CSREES-Administered Higher Education Programs, 160 |
| F-1 | Research, Education, and Economics by Agency for FY 1985–2001 actual, and FY 2002 Estimate, 186 |
| F-2 | Total R&D by Agency, FY 1976–2003, 188 |

| | |
|---|---|
| F-3 | Agricultural Research Funding in the Public and Private Sectors, 1970–1998, 190 |
| F-4 | Amount and Distribution of Major Sources of Revenues of US State Agricultural Experiment Stations, 1980–2000, 192 |
| F-5 | Sources of Revenue for REE Intramural Research Expenditures, 1980–2000, 192 |
| F-6 | REE Agency Funding Allocation by Goal, FY 2000, 194 |
| F-7 | National Summary USDA, State Agricultural Experiment Stations, and Other Institutions, FY 2000, 195 |
| F-8a | Research, Education, and Economics by Function, Agency, and Type of Award for FY 1985–2001 Actual and 2002 Estimate, 196 |
| F-8b | Research, Education, and Economics by Function, Agency, and Type of Award for FY 1985–2001 Actual and 2002 Estimate, 198 |
| F-9 | ARS Funding of Cooperative Activities, 1998–2001, 200 |
| F-10a | Congressional Earmarks for ARS Research and CSREES Research, Education, and Extension, Nominal Dollars, 201 |
| F-10b | Congressional Earmarks for ARS Research and CSREES Research, Education, and Extension, Constant 2000 Dollars, 202 |
| F-11 | Price Index for Research, 2000 Constant Dollar R&D Deflators, 203 |

## FIGURES

| | |
|---|---|
| 1-1 | Growth in agricultural productivity, output, and inputs (1948–1996), page 16 |
| 2-1 | US Department of Agriculture Headquarters Organization, 29 |
| 4-1 | Research, Education, and Economics budget authority by agency for FY 1985–2001 actual and 2002 estimate, 68 |
| 4-2 | Total public and private expenditures, 1970–1998, 70 |
| 4-3 | FY 2000 funding allocation by REE goal, 75 |
| 4-4 | Total CSREES research funding by function, 79 |

## BOXES

| | |
|---|---|
| 2-1 | REE Desired Outcomes and Strategic Objectives, 32 |
| 2-2 | The 22 National Programs of ARS, 34 |
| 3-1 | Research on Relevant Spatial and Temporal Scales, 56 |
| 3-2 | Finding Resources to Explore Research Frontiers, 59 |
| 4-1 | CSREES Competitive-Grant Programs, 81 |
| 4-2 | Examples of REE Responsiveness to Stakeholder Input, 89 |
| 4-3 | Stakeholder Participation and SARE, 90 |
| 5-1 | Research Partnerships in Which REE Has Provided Leadership, 101 |
| 5-2 | Collaborative Activities Through Cooperative Research and Development Agreements (CRADAs), 112 |

6-1   Criteria for Review of NIH and ARS Intramural Research: A Comparison, 126
6-2   Research Position Evaluation System at ARS and the Economist Position Classification System at ERS, 131
6-3   Examples of REE Research Impacts, 139
7-1   The Agricultural Research Service Demonstration Project, 155
7-2   CSREES Investments in Higher Education, 162
G-1   Is ARS Highlighting Its Most Important Research? A Missed Opportunity, 205
G-2   The Carl Hayden Bee Research Center, 206

# Executive Summary

Rapid and dramatic social, economic, and technologic changes have occurred in the food and agricultural sector during the last 30 years. These include increased global competition, the advent of biotechnology and precision production, changes in intellectual property rights, increased product differentiation, greater demand for ecosystem services from agriculture, and changes in farm and market structure. These changes pose new challenges for the federally funded agricultural research, extension, and education system and indicate a need for reflection on the future directions of that system.

In response to a congressional mandate, the US Department of Agriculture (USDA) requested that the National Research Council conduct a study of USDA's Research, Education, and Economics (REE) mission area and provide recommendations for future opportunities and directions. In response to the request, the Research Council convened the Committee on Opportunities in Agriculture and three subcommittees, the Subcommittee on Economic and Social Development in a Global Context, the Subcommittee on Food and Health, and the Subcommittee on Environmental Quality and Natural Resources, which were charged with the following:

1. Drawing in part on previous National Research Council work, the study will include collection, review, and assessment of data on agricultural research and its operating environment.
2. The historical background of agricultural research, education, and economics will be considered, and changes in US needs and priorities will be described.
3. Programmatic and functional complementarities among the four REE

research agencies (the Agricultural Research Service [ARS], the Cooperative State Research, Education, and Extension Service [CSREES], the Economic Research Service [ERS], and the National Agricultural Statistics Service [NASS]) will be examined, and the relevance of agency research to current and proposed national priorities will be evaluated.
4. Current capacity in research, education, and extension will be assessed, and scientific strengths and gaps in federally funded agricultural research efforts will be identified.
5. Research quality will be evaluated for content, relevance, effectiveness, and outcome with regard to the Government Performance and Results Act of 1993.
6. Quality standards, the use of peer review and external advice, resource allocation (including formula funds), and collaborative and interdisciplinary research will be examined.
7. Recommendations will be provided on the future role of federally funded agricultural research, future research opportunities and directions, the setting of relevant research priorities, gaps or weaknesses in the federal agricultural research system, and the strengthening of programmatic, structural, or management components of agricultural research, extension, and education to ensure responsiveness to future national needs.

In responding to its charge, the committee examined the changing context of agricultural research and the widening scope of opportunities for delivering research benefits to society. To capture those opportunities, a renewed federal research enterprise is envisioned, and recommendations are made for changes in research directions, setting of the research strategy, allocation of resources, collaborative relationships, quality and impact assurance, leadership, human capacity, information capacity, and infrastructure.

## THE CHANGING CONTEXT OF AGRICULTURAL RESEARCH

Worldwide changes are transforming American agriculture into an endeavor focused not only on efficient food and fiber production but also on delivering improved public health, social well-being, and a sound environment. Recent scientific breakthroughs will make it easier for agriculture to achieve its potential for delivering a wide array of benefits to society. For this potential to be realized, the agricultural research system must take advantage of new opportunities and relationships and must have the critical leadership in place to address the complex, various roles for agriculture in the 21st century.

Over the last century, the primary public need addressed by US agriculture has been food and fiber production. The major focus of agricultural research, in turn, has been on enhancing agricultural productivity. The success of that endeavor has been substantial, as demonstrated by major productivity gains such as the tripling of corn yields over the last 50 years. Scientific discoveries in such fields as plant and animal genetics, plant and animal nutrition, and livestock health—and effective application of these discoveries in production systems—have driven those gains.

At the same time, important shifts in public values have progressively broadened the scope of agricultural research to include goals related to the environment, human health, and communities. Changing public values and needs will create new market opportunities and will alter agriculture's relationship to the food and fiber system, the environment, and the fabric of American society. The demands for research to support national needs in continued productivity gains, more and varied products, better human health in terms of nutritional outcomes and reductions in foodborne disease, enhanced biosecurity, animal welfare, environmental benefits, and viable rural communities are growing at the same time as scientific advances offer new opportunities for satisfaction of these demands.

## A VISION OF AGRICULTURAL RESEARCH

The changes now under way in agriculture's social and scientific context require a new vision of agricultural research—one that is grounded in lessons from the past, in changing American values, in a globalizing economy, and in scientific advances that have fundamentally altered the life, environmental, and social sciences. The vision promotes agriculture as a positive economic, social, and environmental force. It embraces further gains in food and fiber production—gains that will be crucial to meet the needs of an expanding US and global population—but it also provides other benefits, such as enhanced public health, clean water, more diverse wildlife, rural amenities, and social well-being. In the new vision, agricultural research anticipates the effects of new technologies and emerging socioeconomic structures on society, human health, and the environment. US agricultural research should be conducted with an increased understanding and awareness of how problems and solutions are interconnected globally. International collaboration will be more important for the agricultural research of the future if there is to be real hope of meeting the food and nutrition needs of a growing worldwide population while protecting biodiversity and the environment. Implicit in the vision is a new definition of agriculture's products and consequently of the client base for agricultural research. US agricultural leaders and policy-makers are changing their primary emphasis from production efficiency to meeting changing consumer demands. Food and fiber remain core products but agriculture has an increasingly important role in delivering pharmaceutical, nutritional, and biobased products; the sound stewardship of biologic, land, water, and atmospheric resources; social acceptance of agricultural systems and the well-being of food animals; and the sustained social and economic health of rural communities. The broadening of agriculture's products has greatly expanded the customers of US agricultural research beyond commodity producers. Examples of the new customers are producers of pharmaceutical products; sustainable-, alternative-, and organic-farming interests; a broad array of public and private natural-resource and land managers; conservationists; and entrepreneurs in rural communities.

What kind of federal research enterprise will be required to realize the vision of agricultural research? The enterprise must address a set of priorities in environment, food and health, and social well-being. Better targeting of resources through clear priority-setting mechanisms will improve accountability and make it possible to measure progress against national needs. An emphasis on flexibility will permit targeting of resources and ensure responsiveness to changing public values and rapid advancement of scientific innovations. A system that anticipates challenges arising from emerging technologies, production systems, and consumption patterns—rather than one that simply reacts to problems—will maximize agriculture's long-term benefits. Broad representation of the natural, social, environmental, and health sciences will be essential to reflect the changing portfolio of agriculture's products and the changing client base for agricultural research and to support a multidisciplinary systems approach. The relationships and roles of food and society and more consumer-oriented, health-conscious, global markets should be considered. Partnerships have enormous potential to help further the vision.

**VISION STATEMENT: Agricultural research will support agriculture as a positive economic, social, and environmental force and will help the sector to fulfill ever-evolving demands. These include further gains in food and fiber production and such other benefits as enhanced public health, environmental services, rural amenities, and community well-being. USDA's REE agencies will provide leadership in fostering this concept. Agricultural research will be anticipatory, strategic, collaborative, cost-effective, and accountable to a broad client base. Agricultural research will engage relevant biophysical and socioeconomic disciplines in a systems approach to address new priorities (Chapter 1).**

## FRONTIERS IN AGRICULTURAL RESEARCH

Five challenges provide opportunities for public agricultural research to serve the expanded customer base. For each of those five challenges, we have identified the research frontiers (detailed in Chapter 3) where the intersection of cutting-edge science with stakeholder needs provides compelling opportunities:

- Globalization of the food economy.
  — Evaluate the implications of globalization for US agriculture and agricultural-research priorities.

- — Improve agricultural productivity and product quality while optimizing resource use.
- — Evaluate the economic, social, health, and environmental effects of agricultural technologies and practices.
- Emerging pathogens and other hazards in the food-supply chain.
  - — Reduce the risks of bioterrorism.
  - — Improve microbiologic food safety.
  - — Understand and minimize the hazards of food allergens and toxicants.
  - — Improve understanding and management of plant and animal diseases.
- Enhancing human health through nutrition.
  - — Advance research on bioactive food compounds.
  - — Elucidate genetic mechanisms of human health and nutrition.
  - — Improve understanding of food-consumption behavior and its links to health.
  - — Improve the nutrient content of foods.
- Improving environmental stewardship.
  - — Reduce pollution and conserve natural resources.
  - — Advance environmentally sound alternatives.
  - — Deliver new environmental benefits.
  - — Integrate leading-edge environmental-science concepts and technologies.
- Improving quality of life in rural communities.
  - — Evaluate the effects of changes in agricultural market structure.
  - — Meet the challenge of rural development's changing context.

Research in those frontier fields is often best undertaken in the public sector because many of the challenges will not be fully addressed through private-sector research, inasmuch as the broad environmental and public-health benefits envisioned are widely distributed and cannot be fully captured by private firms. Furthermore, the changing global context for agricultural, food, and rural policies means that USDA policy-makers will require an expanded research base. In many cases, research opportunities will require expanded collaboration among scientific disciplines, federal agencies, or international organizations. Given its historical strengths in mission-oriented research, collaboration on all levels, and responsibility for food and agricultural databases, REE is uniquely positioned to address these frontiers in agricultural research.

**RECOMMENDATION 1: REE should provide leadership for the agricultural community in exploring research frontiers in food, health, environment, and communities. REE should build on its historical strengths and become a scientific leader in using new technologies and emerging scientific paradigms to pursue strategic, long-term research goals. A greater emphasis on multidisciplinary work that engages all relevant disciplines will be needed to address many new research frontiers (Chapter 3).**

REE agencies can build on several strengths in their current capacity and organization as they move into their leadership role. REE has established processes to engage in strategic planning, to ensure quality of science, to listen to stakeholders, to provide for professional development of intramural scientists, and to engage in productive collaborative relationships with a variety of other institutions. Many of those processes can be improved to provide a stronger foundation for leadership.

## SETTING THE RESEARCH STRATEGY

The REE agencies' specific approaches and roles must reflect the changing institutional context of federally supported research. Private-sector expenditures for agricultural research have exceeded public-sector expenditures for 2 decades. USDA-appropriated funds to state agricultural experiment stations are declining, as funding from industry, commodity groups, foundations, and non-USDA federal agencies are providing an increasing share of funding. USDA remains an important source of funding for most agriculture-related research, but it faces increased challenges in providing leadership for agricultural R&D and must be more strategic in use of its funds and in articulating the importance of outcomes of agricultural research.

Federal resources should be used to support outcomes with broad public benefit that are not well funded by private-sector interests. Such benefits as enhanced public health or environmental services are often more difficult for private firms to capture, so these are important goals for public research. Federal investments also include research that provides new "platforms" of discovery for multiple private or local applications.

**RECOMMENDATION 2: The REE agencies need to identify clearly their unique positions relative to the other components of the agricultural-research system, identify high-impact activities through which targeted funding and resources could generate substantial and measurable progress toward meeting national needs, and coordinate planning and research support across the agencies to minimize unnecessary duplication and maximize effectiveness. Those efforts should be informed by a clear articulation of the major national priorities for research and education and a system for anticipating, reporting on, and identifying strategies to address emerging research needs (Chapter 4).**

### Resource Allocation

REE has substantial resources invested across the broad array of research goals related to agriculture, food, health, environment, and communities. However, agricultural productivity still receives the dominant share of research resources, particularly for intramural research. Without question, the REE research agenda of the future will require greater resources and a more balanced distribution of those resources in new directions mentioned earlier.

**RECOMMENDATION 3: The REE agencies should direct new and existing resources that currently support agricultural productivity research toward new research opportunities in health, environment, and communities (Chapter 4).** (Research opportunities are identified in Chapter 3.)

Approaching the research frontiers requires new resource-allocation strategies. The four current funding mechanisms in agricultural research are formula funding, competitive grants, special grants or earmarks, and intramural research. The diversity of financial sources usually ensures that local, state, regional, and national agricultural research needs are addressed, and economic evidence suggests that the diversity of funding mechanisms has been a historical strength of the USDA research system. It is unclear, however, what role competitive grants will play in the overall research portfolio, given their variability in appropriated funds in the last few years. It is also unclear whether the current portfolio of funding mechanisms will adequately address the complex problems of contemporary agriculture in the 21st century and realize the new vision of REE research. In particular, additional flexibility is needed to help the REE agencies respond most effectively to opportunities and to help provide the research results that are needed by USDA agencies administering programs mandated by the department (action agencies).

A realignment of the existing research budget to increase the proportion of funds in competitive grants and cooperative agreements would be effective in achieving greater flexibility and for addressing new and emerging issues by engaging new talent and expertise. A target of 20–30% of the research portfolio allocated via merit-reviewed competitive processes would achieve greater parity with other federal research programs. The committee envisions that increasing the percentage of research allocated via cooperative agreements to 25% of the portfolio would also contribute to greater flexibility. Competitive mechanisms should be used for awarding large cooperative agreements ($1 million or more). However, competitive mechanisms need not be used in the case of small awards for which competitive processes may incur delays and higher transaction costs. Greater discretion for REE agencies to move resources to new subjects could also be achieved through no-year-funding or revolving-funding authority or by withholding a percentage of discretionary funds for research in new subjects. Discretionary funds withheld above the agency level could be used as an incentive for agency collaboration on emerging issues or emergency needs.

The committee believes that allocating discretionary resources for research to action agencies could also contribute to more effectively meeting action-agency research needs. REE would be well placed to receive these resources, but a more competitive mechanism would create greater accountability and transparency in terms of carrying out research designed to meet the needs of action agencies.

**RECOMMENDATION 4:** To ensure that research funds are used to advance science in new directions and to address emerging and emergency issues in a timely and responsible fashion, the committee recommends the following (Chapter 4):

1. Total competitive grants should be substantially increased to and sustained at 20–30% of the total portfolio.
2. Action agencies should receive or control discretionary funds to be used to meet critical programmatic needs complementary to those currently served by REE agencies. The agencies could thereby fund intramural USDA scientists, other agency scientists, or university researchers competitively on the basis of the researchers' availability and match of expertise to agency needs.
3. The REE agencies should pursue complementary research activities and tap broader expertise by dedicating a higher percentage of new funds to cooperative arrangements, to be awarded on a competitive basis for large awards, with academic or other public-sector researchers.
4. Congress should increase REE budgetary flexibility to move resources toward emerging and emergency needs.

### Relevance Assurance Through Stakeholder Input

The REE agencies have implemented numerous mechanisms to integrate stakeholder input into their priority-setting and into the research, extension, and education processes. Stakeholder input generally strengthens the connection between research and its applications, but results of REE's efforts to engage stakeholders have been mixed. Not all processes have ensured balanced participation by the full array of affected stakeholders. Efforts have been largely unlinked across agencies, and this creates duplication of effort and sometimes disparate results. The stakeholder processes now occurring strain stakeholder time and resources and the capacity and resources of the REE agencies. These processes do not effectively utilize the national cooperative extension network that exists at local and state levels. Moreover, the agencies have sometimes found it difficult to reconcile stakeholders' competing views and to synthesize diverse and abundant stakeholder input into a usable form. Finally, stakeholder processes are as yet weakly linked to REE and its agencies' strategic planning and performance evaluation.

**RECOMMENDATION 5:** To provide a forum for shared learning across agencies, REE should conduct a national summit every 2–3 years that would engage the four REE agencies and a broad representation of stakeholders at the local, national, and regional levels. The summit could assess national research needs and inform stakeholders how their input is used in agency decision-making (Chapter 4).

## COLLABORATION

Partnerships between REE and universities over the last 50 years have worked effectively in addressing many of agriculture's greatest challenges, such as soil conservation. The emergence of new kinds of research organizations and structures is now providing opportunities for REE to explore different kinds of partnerships and research collaborations and is challenging conventional ways of carrying out research. Examples of emerging and continuing partners in REE research are other federal research agencies involved in human health and the environment, nonprofit organizations, international research centers, and agricultural research systems in other countries. All those kinds of partnerships can play important roles in addressing new research opportunities.

Collaborations with the private sector are growing rapidly. Policy changes over the last 2 decades allow patenting and licensing of knowledge developed through public-sector research (for example, the Bayh-Dole Act) and have expanded the scope of collaboration between the public and private sectors, opening new opportunities and risks in technology development. Benefits of such public–private collaboration include more-successful technology transfer, increased support of research, and expanded scientific networks. Concerns about such collaboration include its potential effect on priority-setting in the public sector, on scientific-information generation, and on the allocation of resources for future research. Many questions regarding the management of intellectual property in agriculture are unresolved, and policy is still not well defined.

**RECOMMENDATION 6: REE should provide national leadership in developing intellectual property policy for agricultural research. REE should address the potential consequences of public–private collaboration with appropriate policies, practices, and organizational arrangements that**

- **Promote the greatest public benefit from agricultural research.**
- **Protect the public investment in research.**
- **Prevent diversion of public resources away from research that can be carried out only in the public sector.**
- **Pursue strategic private-sector collaboration necessary to achieve public goals.**

To accomplish these objectives, REE should establish ways to measure the effectiveness of technology generation and transfer through private-sector collaboration (Chapter 5).

## QUALITY AND IMPACT ASSURANCE

The committee considered the REE agencies in light of metrics of quality and mechanisms for quality assurance. Generally, the committee identified a variety of quality-review and evaluation processes that are in place for all research projects and programs in the REE system. The committee found evidence that REE scientists produce research of high quality. The adoption of peer-review systems in both the intramural and formula-funded research systems is a positive step. However, in comparing REE with other federal research programs, the committee found that the REE system appears to reward excellent research performance adequately but—except for the competitive programs—may not adequately exclude mediocre research performance. In benchmarking REE against other federal intramural research programs, the committee found that unsatisfactory performance has very little consequence in the REE intramural system, whereas in some other federal intramural research programs, reduction or complete loss of research support or ineligibility for tenure was a consequence of unsatisfactory research performance.

**RECOMMENDATION 7: The REE intramural research system should strengthen quality control for poor research performance. Mechanisms used in other federal intramural research agencies, including the redirection of human or financial resources when quality is poor, could be implemented (Chapter 6).**

The committee reviewed the impact of REE research by using a variety of metrics focused on some dimension of the real output or payoff of research, including publications, citation frequency, patenting, longer-term quantitative measures, and social rate of return. In general, the impact of REE research on several important outcomes is well documented. The social rate of return on past public agricultural-research investments in the period 1950–1982 has been very high, and the rate of return over the last 2 decades has not declined. Although documented quantitative data are not as systematized as economic rates of return, there is evidence that REE research has had positive environmental, social, nutritional, and health impacts.

### Monitoring and Communicating Impact

The new research agenda poses challenges for tracking and monitoring success. Public investments in agricultural research have shown a high rate of return over the last 75 years, primarily through enhanced productivity. But future investments will yield improvements in public health, the environment, and community well-being that are more difficult to measure. Tracking outcomes and measuring success at these research frontiers require strategic thinking about information collection.

**RECOMMENDATION 8: REE agencies should develop and adopt ways of measuring the national, long-term impacts of their research on the environment, human health, and communities. The tools should include measures and indicators that are influenced by agricultural research or that can be attributed to research outcomes, including how research supports the needs of action agencies. REE should strive to achieve greater transparency in communicating these impacts through timely electronic publishing of peer-reviewed results and through greater efforts to interpret these results for a general audience (Chapter 6).**

Monitoring capability should be developed to show how REE research is changing in focus, relevance, quality, leadership, and accountability. Monitoring capability should also be developed to show how food, agricultural, natural, and human systems are changing with a view toward targeting future research directions. More effective tracking capability will help to improve self-evaluation in REE agencies and reporting of progress to groups outside REE.

## REE CAPACITY

The dramatic changes in science and technology, globalization, emerging needs, and the identification of new research themes commensurate with a broader scope of societal issues will require changes in leadership, scientific staffing, and data development and management.

### Organizational Capacity and Research Leadership

The current organizational structure of research efforts in the REE agencies limits the combined effectiveness of the agencies. Leadership to provide intellectual guidance and a long-term, coherent vision for REE research, promote intra-agency coordination, broker partnerships outside the REE agencies, and integrate REE's research within the federal research program is lacking. No position in the REE administrative structure has the visibility and prestige of the directors of the National Institutes of Health and the National Science Foundation, and the scientific reputation of the REE agencies suffers from this lack.

**RECOMMENDATION 9: There is a national need for a high-level leader to represent food and agricultural research and to promote opportunities for the research system. Such a leader should be vested with the**

authority to develop the food and agricultural research agenda, redirect funds to emerging issues and emergency needs, integrate the efforts of the individual agencies, and facilitate collaboration and coordination with scientists outside USDA and elsewhere in the federally supported research system. The leader should be selected on the basis of outstanding scientific and administrative accomplishments and must command the respect of the agricultural community and the broad scientific community (Chapter 7).

Most committee members believe that creating a new position of research director who reports directly to the secretary of agriculture would be the best among several alternatives for establishing the high-profile leadership that is needed to implement the new vision for food and agricultural research described in this report. Several committee members concluded that other options, including strengthening the undersecretary position, also could successfully address the need for enhanced leadership of the nation's food and agricultural research effort.

## Human Capacity

Evidence suggests that the scientific expertise needed to address progressive fields of research is lacking in REE. There is a continuing lack of scientific expertise in the nutritional, environmental, and social sciences and imbalances in ethnicity and sex within and between agencies.

**RECOMMENDATION 10: REE should increase the hiring of scientists in research fields that have the greatest opportunities to address societal goals. Those include integrative environmental science, ecology, economics, and sociology; human genetics (including statistical human genetics) and bioinformatics; and human nutrition, public health, and food safety. REE agencies should continue to develop new methods for recruiting and retaining women and members of ethnic minorities (Chapter 7).**

The committee found that REE agencies face a number of recruitment and retention challenges, including stiff competition from the private sector, other federal agencies, and academe, complex and constraining hiring rules under the Office of Personnel Management, and an increasing number of non-US citizens with PhDs in agricultural sciences who are not eligible for employment in US government agencies. In spite of these challenges, REE agencies have made use of a variety of recruitment procedures and retention incentives to increase flexibility. These include the postdoctoral fellowship program, the Demonstration Project, cooperative agreements with university-based research centers, the recently authorized Senior Scientific Research Service program, and incentives authorized under the Federal Employees Pay Comparability Act.

The committee also notes that REE has made substantial efforts to build internal capacity by promoting training and professional development. REE also contributes to building the human capacity of the future through its support of the research establishment in the land-grant university system. Continued efforts of these kinds hold promise for the future.

## Information Capacity

A broader perspective in surveying and in collecting data will be necessary to support an expanding and broadening food and agricultural research agenda, particularly in the areas of nutritional and environmental analysis. New categories of data and systems for data use are needed. Creative approaches to data collection and analysis that integrate the unique strengths and complementary expertise of all the REE agencies, land-grant universities, other government agencies, the private sector, nongovernment and voluntary groups, and international organizations should be implemented. Finally, new technologic tools, including geospatial referencing, are enabling the combination of new and existing datasets from different sources to create new knowledge.

**RECOMMENDATION 11: REE should undertake an analysis of the data development, management, and dissemination needed to support environmental and nutrition policy analysis. REE should work with other USDA mission areas to conduct an inventory of available social, economic, biologic, chemical, and physical datasets and to take stock of the data needs of the future. REE should take the initiative in coordinating with other USDA agencies and with other federal agencies to identify where and how data can be more efficiently and effectively used and shared. REE should put into place structures and systems to support data management and dissemination across its agencies (Chapter 7).**

## Research Infrastructure

State-of-the-art facilities and equipment are critical requirements for USDA to be able to conduct world-class science and research. However, maintaining a physical infrastructure that is too large and too expensive will have a major adverse effect on department research unless REE budgets grow substantially or REE is able to gain in efficiency by being permitted to close and consolidate a number of facilities. Maintenance of some facilities has been deferred for many years, and the cost to repair these facilities is mounting to tremendous sums of public funds. Congressional and stakeholder pressures greatly hinder the ability to close some facilities that do not cost-effectively contribute to USDA's national research agenda.

**RECOMMENDATION 12:** The committee recommends that REE use objective criteria to decide which USDA facilities merit investment of budget resources for repair, modernization, or security improvement and which should be consolidated or closed because they are incapable of cost-effectively contributing to the REE research strategy without renovation. These criteria should be established in the public interest and mutually agreed on by key members of Congress and state and local legislators, as articulated in the principles and recommendations of the 1999 Report of the Strategic Planning Task Force on USDA Research Facilities. The closing, consolidation, or renovation of facilities should be implemented (Chapter 7).

## LOOKING TO THE FUTURE

As the world's premier agricultural research system, USDA and its partners have been widely emulated. The increasingly international character of research benefits means that USDA's future choices will have global consequences. Partners in the research effort will be increasingly diverse and far-flung, and how USDA chooses to partner with other institutions will provide models for global collaboration. USDA can lead the way for institutional change that responds to new demands on the agricultural system.

# 1

# Vision and Leadership

Worldwide changes are transforming American agriculture into an endeavor focused not only on efficient food and fiber production but also on improving public health, social well-being, and the environment. Recent scientific breakthroughs will make it easier for agriculture to achieve its potential for delivering a wide array of benefits to society. But for that vision to be realized, the agricultural research system must take advantage of new opportunities and new partnerships and must have the leadership to address the complex and varied roles of agriculture in the 21st century.

Over the last century, the primary public need addressed by US agriculture has been food and fiber production, and the major focus of agricultural research has been on maximizing the productivity of agronomically important crops and livestock. The success of that endeavor has been substantial as demonstrated by such productivity gains as the tripling of corn yields over the last 50 years (USDA, 2002b) and an increase in overall productivity by 2.5 times during the last 50 years (Figure 1-1; USDA, 2000) and by the low average percentage (10.2%) of consumer income spent on food in the United States (USDA, 2002a). Scientific discoveries in plant and animal genetics, plant and animal nutrition, and livestock health—and effective application of these discoveries in production systems—have driven those gains.

At the same time, important shifts in public attitudes have broadened the scope of agricultural research to include goals related to the environment, human health, and communities. Changing public attitudes and needs will create new market opportunities and will alter agriculture's relationship to the food and fiber system, the environment, and the fabric of American society. The increasing pace of scientific discovery and technology development will revolutionize

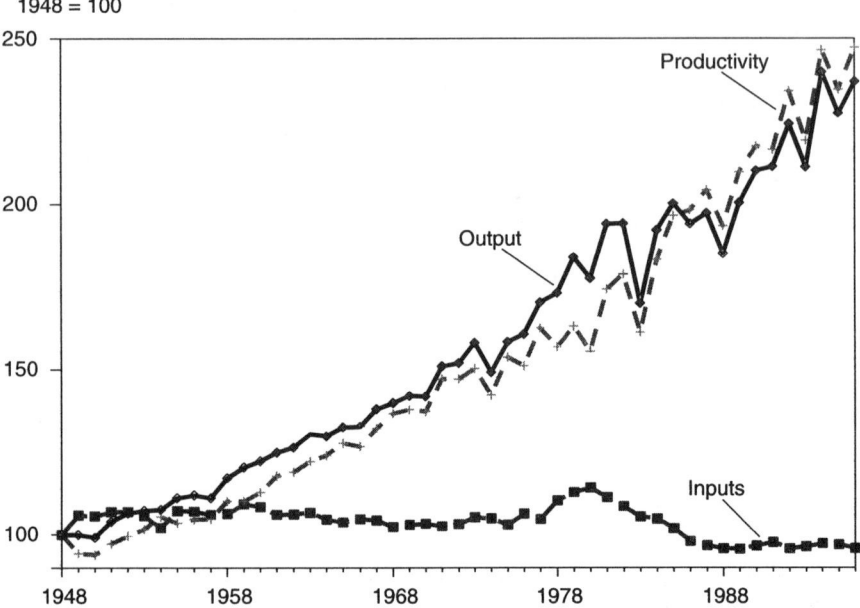

**FIGURE 1-1** Growth in agricultural productivity, output, and inputs, 1948–1996. Source: USDA (US Department of Agriculture). 2000. Agricultural Resources and Environmental Indicators. Washington, DC: Economic Research Service, US Department of Agriculture. Available online at *http://www.ers.usda.gov/Emphases/Harmony/issues/arei2000/*.

agriculture's capabilities. We identify here some key changes unfolding today, their implications for the direction of research administered by the Research, Education, and Economics (REE) agencies of the US Department of Agriculture (USDA), and the need for strong leadership to manage and lead change effectively.

## CHANGING PUBLIC ATTITUDES AND NEEDS

Participants in American agriculture now operate in a highly competitive global economy. Globalization has changed the nature of agricultural products and the system that produces them. Trade liberalization provides great opportunities for expanding US agricultural markets overseas, but it also allows aggressive competition from overseas producers. There is increased public sensitivity to and awareness of global social and economic challenges, including population growth, food insecurity, and poverty. Operating in this competitive environment will require greater flexibility, improved management and decision-making, and continued advances in agricultural productivity. A key challenge for agricultural research will be to balance the continued need for productivity and efficiency gains with emerging demands for new products and for environmental and social services. There is continued tension between a realignment of agriculture's benefits in food safety, nutrition, conservation, and so on, and primary incentives to continue increasing production, sometimes at the expense of other priorities. That is clear from the large increases in agricultural subsidies in the 2002 farm bill (US Congress, 2002). As discussed in Chapter 4, the fact that the actual budget distribution across program areas does not align with stated objectives for budget distribution is consistent with the lack of incentives to move from the status quo.

The number and diversity of products yielded by the global agricultural system are expanding rapidly, and the products now include pharmaceuticals and other health-promoting foods. Even within the traditional food and fiber sector, more items are sold today than ever before, including an increasing number of value-added products. Changes in consumer preferences related to shifts in demography, affluence, global demand, and education levels account in part for that trend. Consumer acceptance of "functional" foods—foods whose components are associated with good health and decreased disease risk and include dietary supplements or "nutraceuticals"—has made such foods the subject of a significant trend in the food industry (Childs, 2001). The result is a convergence of the global food and pharmaceutical industries that is creating a new "agriceutical" industry composed of multinational public and private entities and focused on human health and nutrition.

Never before has the linkage between agriculture and public health been more apparent, vital, or promising. The new research agenda will need to expand its role and resources to take advantage of this unprecedented opportunity. For example, the growing public interest in food safety reflects awareness of this linkage (Unnevehr and Roberts, 2002). The incidence of foodborne illnesses in America is rising. The increased frequency of eating away from home (USDA, 2001b) and changing food-consumption patterns have enabled the emergence of and exposure to new pathogens. In addition, an increasing percentage of the population is becoming susceptible to opportunistic infections, including foodborne pathogens, given the rising percentage of the US population over 65, a growing number of persons infected with HIV, and the growing numbers of recipients of bone marrow or organ transplants and patients receiving chemotherapy or immunosuppressive drugs (CAST, 1994; USDHHS, 1998). Increased movements of animals, people, and products are introducing new and unfamiliar risks into the food system. Epidemiologic evidence suggests that some kinds of animal production systems—including operations with higher animal densities and mechanization systems that disperse feeds, water, and other inputs and outputs—may increase human exposure to infectious disease. About 75% of new human pathogens over the last few decades have originated in or been transferred through livestock, poultry, and wildlife, including the bovine spongiform encephalopathy prion, *Salmonella enteritidis*, and *Escherichia coli* O157:H7, demonstrating the continued importance of animal sources in the transmission of so-called emerging pathogens (Tauxe, 1997).

Public sensitivity about food safety is particularly high because of national concerns about terrorism, the nation's food security, and the vulnerability of our agricultural resources. That sensitivity fits within a larger trend of greater public interest in food origins. Several well-publicized outbreaks of foodborne pathogens have highlighted questions about disease sources and mitigation. Identification in human food products of genetically modified corn not yet approved for human consumption (US Congress, 2001) has raised concerns about the traceability of and accountability for food origins. Those and other issues have helped to fuel the rapid expansion of consumer demand for organic products and products from low-input agricultural systems over the last decade.

Another important transformation is under way in how American society views the relationship between agriculture and the environment. Numerous public policies enacted over the second half of the 20th century sought to reduce the harmful environmental effects of agricultural intensification and widespread pesticide and fertilizer use. Today, however, the public is asking agriculture to go further and to deliver environmental benefits. That trend began with establishment of the Conservation Reserve Program and the Wetlands Reserve Program and continued in recent discussions on the conservation title of the farm bill (US Congress, 2002). The lands are expected to play an increasingly important role in providing clean water, mitigating global climate change, conserving the world's biologic diversity, and maintaining rural amenities, such as open space and recreational opportunities. Indeed, national demand for environmental and recreational services from the land is expected to outstrip demand for food in some areas, much as the recreational value of many national forests now exceeds their timber value (Sedjo, 1998). There is also increased public awareness of and concern about global environmental change and challenges, including natural-resource degradation, desertification, climate change, and loss of global biodiversity.

There have been important changes in the relationship between agriculture and rural communities. Agricultural production has become highly concentrated among fewer and fewer farms over the last century (NRC, 2002) and among larger operations. Farmers with annual gross sales of more than $250,000, representing 8% of US farmers, produced 68% of the nation's agricultural production in 1999. US farmers with annual gross sales of less than $250,000, representing 92% of US farmers, produced only 32% of total agricultural production in 1999 (USDA, 1999). The farm population is quite diverse in economic circumstances and in sources of income. In comparison with the general US population, inequality in household income is greater among farm households (Lobao, 1990; Lobao and Meyer, 2001; Mishra et al., 2002; USDA, 2001c). Agricultural decision-making and the adoption of new technologies increasingly involve relatively few large producers (NRC, 2002). These distributional differences are associated with a decline in the social and economic vitality of many rural communities and persistent poverty in some areas despite numerous incentive and investment programs designed to reinvigorate rural economies. Agriculture has become a much smaller part of the rural economic base; farming is the primary economic activity of only one-fourth of rural counties[1] (USDA, 1994). Persons living on farms constituted only 5.1% of the rural population in the 2000 census (USDC, 2002). Farm production and closely related employment[2] accounts for 12.5% of the total rural employment (USDA, 2001d). Agricultural productivity itself cannot ensure the economic health of rural communities, pointing to a need for new opportunities.

## RECENT INNOVATIONS IN SCIENCE AND TECHNOLOGY

The last few decades have seen advances across the spectrum of the life sciences and social sciences—from molecular biology to ecosystem dynamics. Technical innovations resulting from those advances have begun to alter the practices and products of agriculture fundamentally. The availability of new tools in turn provides further opportunities for research.

### Biotechnology and Genomics

Beginning in 1983, scientists have introduced novel gene sequences into plants to confer resistance to specific insects, viruses, and herbicides; by 1995, transgenic crops that carry resistance traits were in commercial production. Transgenic varieties of cotton, corn, soybeans, tomato, squash, and papaya have fundamentally altered how seeds, crops, and foods are developed, produced, sold, and regulated.

Current studies in plant genomic sciences promise to provide additional breakthroughs that will influence how future crop varieties are developed. Genetic mapping techniques that use DNA markers are increasing the rate of breeding of new crop varieties. Modern techniques for isolating and characterizing genes and for determining the function of genes have led to an astounding leap in knowledge. Scientists have identified genes that are involved in cold, drought, and saline tolerance; genes that control flowering and vegetative growth; genes that control reproductive functions and embryo development; genes that confer resistance to fungi, bacteria, nematodes, and viruses; and genes that con-

---

[1] In farming-dependent counties, farming contributed a weighted annual average of 20% or more of the total labor and proprietor income in 1987–1989 (USDA, 1994).

[2] Closely related employment includes agricultural services, agricultural input industries, and agricultural processing and marketing (USDA, 2001d).

trol levels of plant hormones and secondary metabolites that can impact human health and nutrition. The discoveries have occurred in crop plants, as well as model plants, and future work promises to deliver those and other traits to crop plants and to create ever greater opportunities for agriculture to affect human health and nutrition, sustainability in crop production, and crop productivity.

Genomics tools are being used to study the genetic predisposition to environmental influences leading to human and animal disease. The tools will also be used to describe the impact of the chemical components of foods on disease conditions and will lead to a better understanding of the links between human health and nutrition. Increasingly, collaborations between nutritionists and researchers in the health sciences with plant scientists will create opportunities to develop foods that mitigate diseases and predispositions to diseases.

Advances in nutrition science in the 1990s expanded understanding of essential nutrients and their role in the etiology of major diseases. That set the stage not only for recent growth in "functional" foods with specific nutritional attributes but also for future development of nutritionally fortified foods through biotechnology. Advances in our understanding of animal nutrition and genetics have resulted in major gains in efficiency and quality in the dairy, livestock, poultry, and pork industries that are expected to enhance the future competitiveness of US animal agriculture. As the cloning of farm animals develops to commercial use, animal feeds are expected to be developed to match the genetics of the animals, and this should lead to more efficient growth and meat production, increased compatibility of meat with human dietary needs, and reduced waste and environmental pollution from animal production facilities. Advances in disease detection and control, including incorporation of vaccines and other preventives in feeds, will reduce the bacterial, fungal, and viral contamination of animal products, further increasing production efficiency and food safety.

## Ecosystem and Social Dynamics

A more sophisticated understanding of the spatial and temporal dynamics of ecosystem patterns and processes has led to the emergence of new disciplines, including agricultural ecology, landscape ecology, ecosystem management, and earth-system science. The coupling of concepts from the new disciplines with new analytic frameworks and spatial technologies, such as geographic information systems and global positioning systems, is yielding powerful tools for understanding the interactions between agricultural practices and the functioning of adjacent and distant ecosystems. The advances in ecology are revealing that such interactions are far more complex and far-reaching than previously thought. The environmental benefits or harmful effects of some agricultural practices can be additive or multiplicative and even be seen to change qualitatively when viewed over increasingly large spatial scales, over a greater diversity of ecologic systems, or over extended periods. Global-change processes related to climate, nitrogen deposition, and land use that are now being documented on very large spatial scales will have profound implications for the global environment and are partly the result of actions on many individual farms. Yes substantial gaps exist in our understanding of these interactions and therefore of how the actions of individual farmers might be adjusted to help mitigate global environmental problems.

Global data on natural resources and the technologies for managing, manipulating, and applying this information are evolving rapidly, enabling the testing of hypotheses that could not previously be tested. Tools being developed will integrate spatially referenced and satellite-based, remotely sensed data into decision-support systems for farms, forests, and rangelands. Large new databases have provided the raw material for improved epidemiologic approaches for understanding, preventing, and minimizing disease outbreaks. Transfer and manipulation of massive datasets among researchers have become routine. And simultaneous access to multiple databases through the Internet has enabled synthetic data analyses that previously were impossible.

An equally sophisticated understanding of the social and economic interactions between farm and nonfarm sectors has emerged through advances in the social sciences. For example, new analytic and modeling methods have made it possible to test the impacts of competing policy options in addressing a broad set of social goals. Burgeoning information resources are allowing analyses of demographic, economic, and environmental effects of trade and immigration trends. Emerging scientific approaches for exploring the interplay of social and biophysical processes—for example, modeling approaches for assessing how changing economic conditions affect land-use decisions and ecologic conditions—are expected to yield important insights into the determinants of environmental quality and the effectiveness of various policy approaches (e.g., Costanza, 1995; Matson et al., 1997; NRC, 1999; Parks, 1991; Sengupta et al., 2000).

The social and communication sciences have created a new human dimension for understanding food safety and the acceptance of foods. The appreciation of risk assessment, risk communication, consumer education, and human behavior and attitudes are examples of the blending of biomedical and social sciences. The advent of genetically modified crops and animals has added to the importance of the human dimensions of contemporary agriculture and related research.

## A VISION FOR THE FUTURE

The changes now under way in agriculture's social and scientific context require a new vision of agricultural research—one that is grounded in lessons from the past, in changing American values, in global changes and challenges, and in scientific advances that have fundamentally altered the life, environmental, and social sciences. The new vision promotes agriculture as a beneficial economic, social, and environmental force. It embraces further gains in food and fiber production—gains that will be crucial to meet the needs of an expanding US and global population—and it provides other benefits, such as enhanced public health, clean water, wildlife, rural amenities, and social well-being. In the new vision, agricultural research anticipates the effects of new technologies and emerging socioeconomic structures on society, human health, and the environment. Agricultural research is much more global in scope and consideration than in the past. The success of USDA's future agricultural research will be determined by how it adapts to and manages change, innovation, entrepreneurism, and by a change in culture in how USDA research agencies work, with whom they work, and what they will work on. This is an unprecedented time in the history of agricultural research and a time in which there is a special premium on strong leadership skills (discussed in Chapters 4 and 7).

Implicit in the new vision and the need for leadership is a new definition of agriculture's products and thus of agricultural research's client base. US agricultural leaders and policy-makers are changing their primary emphasis from production efficiency to meeting changing consumer demands (ESCOP, 2001; USDA, 2001a). Food and fiber remain core products, but agriculture has an increasingly important role in the delivery of pharmaceutical, nutritional, and other biobased products; the sound stewardship of biologic, land, water, and atmospheric resources; the well-being of food animals; and in continuing to sustain the social and economic health of rural communities. Just as agricultural producers of the future will have an expanded role as global marketers and as environmental stewards, they will also need to be strong public-health advocates. As food and health are being linked in new ways, producers are being linked more closely with consumers, and agricultural products with human health, well-being, and productivity. The broadening of agriculture's products has greatly expanded the customers of US agricultural research results beyond commodity producers. Examples of the new customers are producers of pharmaceutical products; sustainable-, alternative-, and organic-farming interests; a broad array of public and private natural-resource and land managers; conservationists; rural communities; and government agencies. (Mechanisms for ensuring the relevance of research to stakeholder needs are discussed in Chapter 4.)

What kind of federal research enterprise will be required to realize the new vision of agricultural research? It must address a new set of priorities in environment, food and health, and community well-being (discussed in Chapter 3). The research enterprise must reconsider food and society and their new relationships and roles and must shift its emphasis to consumer-oriented, health-conscious, global markets. Better targeting of resources through clear priority-setting mechanisms will improve accountability and make it possible to measure progress against national needs (discussed in Chapter 4). An emphasis on flexibility will ensure responsiveness to changing public values and rapid development of scientific innovations. (Funding mechanisms that contribute to greater flexibility are discussed in Chapter 4.) A system that anticipates challenges arising from emerging technologies, production systems, and consumption patterns—rather than one that simply reacts to problems—will lead to larger long-term net benefits for agriculture. Agriculture is a system that links many physical, biologic, social, and economic processes. Tomorrow's agricultural research must explicitly identify and address these linkages so that progress in one agricultural sector does not inadvertently create or exacerbate problems in another sector. Broad representation of the natural, social, environmental, and health sciences and consideration of relevant temporal and spatial scales will be essential to reflect the changing portfolio of agriculture's products and the expanding client base of agricultural research and also to support a multidisciplinary, systems approach (discussed in Chapters 5 and 7).

The REE agencies' specific approaches and roles must reflect the changing institutional context of federally supported research. REE funding today is a minor component of overall US funding of agricultural research, given the increasing contribution of state governments, industry, and other federal agencies (see Chapter 4). Consequently, REE resources, always limited, should be targeted at efforts in which they can make a unique, critical, and high-impact contribution to the public good. One such effort is the response to major national needs identified in Chapter 3, which are outcomes of the changing context for agriculture described above. Within these national needs, federal research must increasingly focus on basic research to create new platforms for private applications, which may often include long-term projects that could not exist on shorter time horizons. Federal research must also be directed toward outcomes with positive spillover benefits for the environment and public health. Federally supported research would thus complement, not duplicate, the emphasis of research funded by the private sector.

Partnerships between REE agencies and universities over the last 50 years have been effective in addressing many of agriculture's greatest challenges, such as soil conservation. The emergence of new kinds of research organizations and structures is now providing opportunities for REE agencies to explore different kinds of partnerships and research collaborations at the same time as it challenges conventional ways of carrying out research. Policy changes allowing patenting and licensing of products of publicly funded research (such as the Government Patent Policy Act of 1980 [US Congress, 1980]) have expanded the scope of collaboration between the public and private sectors, opening new opportunities and risks in technology development. The new breed of potential USDA partners also includes nonprofit research institutions, public-interest groups, and other federal agencies involved in human health and the environment. New and more effective partnerships must be solidified among the USDA agencies, the National Institutes of Health, the Food and Drug Administration, the Environmental Protection Agency, and other federal agencies. REE collaboration with international partners will be even more important in the future in contributing to solving global challenges. Scientists have only begun to glimpse how sophisticated information technologies will revolutionize research relationships. Networked "virtual laboratories" already enable researchers separated by miles and even continents to collaborate on shared ideas, data, and manuscripts; they provide a powerful new tool for supporting the multidisciplinary work that will be increasingly important for REE agencies. USDA researchers will need to engage and encourage new voices in their decision-making and priority-setting. (Collaboration and new partnerships are discussed in Chapter 5.)

To address a broader set of research goals and to do so with greater accountability, flexibility, foresight, and collaboration is a substantial challenge for the REE agencies. There are a variety of structural and cultural obstacles to change, including narrowness in scope, narrowness in discipline, insularity in style and approach, and resistance to change. Strong leadership will be necessary to surmount these obstacles and to achieve the vision (discussed in Chapter 7). The body of this report identifies some of the key research opportunities that lie ahead for the REE agencies and some of the institutional and cultural changes that will enable USDA to realize the new vision of agricultural research.

**VISION STATEMENT: Agricultural research will support agriculture as a positive economic, social, and environmental force and will help the sector to fulfill ever-evolving demands. These include further gains in food and fiber production and such other benefits as enhanced public health, environmental services, rural amenities, and community well-being. USDA's REE agencies will provide leadership in fostering this concept. Agricultural research will be anticipatory, strategic, collaborative, cost-effective, and accountable to a broad client base. Agricultural research will engage relevant biophysical and socioeconomic disciplines in a systems approach to address new priorities.**

## SUMMARY

This chapter has offered a vision for agricultural research in context of advancing science and technology and changing public attitudes and needs. Globalization, trade liberalization, changes in consumer preferences, public concern about food safety and the environment, and changes in the relationship between agriculture and rural communities have altered the context in which agricultural research is conducted. Emerging approaches in biotechnology and genomics, ecosystem science, and social science have also transformed the practices and products of agriculture and have provided new opportunities for research. Agricultural research that holds promise for new benefits in public health, the environment, rural amenities, and community well-being and is anticipatory, strategic, collaborative, cost-effective, and accountable to a broad client base is envisioned.

# REFERENCES

CAST (Council for Agricultural Science and Technology). 1994. Foodborne Pathogens: Risks and Consequences, Task Force Report No. 122, September, pp. 23–24.

Childs, N.M. 2001. Marketing issues for functional foods and nutraceuticals. In Handbook of Nutraceuticals and Functional Foods, R.C. Wildman, ed. New York: CRC Press.

Costanza, R. 1995. Ecological economics: Toward a new transdisciplinary science. Pp. 323–348 in A New Century for Natural Resources Management, R.L. Knight and S.F. Bates, eds. Washington, DC: Island Press.

ESCOP (Experiment Station Committee on Organization and Policy). 2001. A Science Roadmap for Agriculture. Washington, DC: National Association of State Universities and Land-Grant Colleges. Available online at *http://www.nasulgc.org/publications/agriculture/science%20Roadmap2.pdf*.

Lobao, L. 1990. Locality and Inequality: Farm and Industry Structure and Socioeconomic Conditions. Albany, NY: State University of New York Press.

Lobao, L., and K. Meyer. 2001. The great agricultural transition: Crisis, change, and social consequences of twentieth century US farming. Annual Review of Sociology 27:103–124.

Matson, P.A., W.J. Parton, A.G. Power, and M.J. Swift. 1997. Agricultural intensification and ecosystem properties. Science 277:504–509.

Mishra, A.K., H.S. El-Osta, M.J. Morehart, J.D. Johnson, and J.W. Hopkins. 2002. Income, Wealth, and the Economic Well-being of Farm Households. ERS Agricultural Economic Report No. AER 812. 77 pp. July. Washington, DC: Economic Research Service, US Department of Agriculture. Available online at *http://www.ers.usda.gov/publications/aer812/*.

NRC (National Research Council). 1999. Our Common Journey: A Transition Toward Sustainability. Washington, DC: National Academy Press.

NRC (National Research Council). 2002. Publicly Funded Agricultural Research and the Changing Structure of US Agriculture. Washington, DC: National Academy Press.

Parks, P.J. 1991. Models of forested and agricultural landscapes: Integrated economics. Pp. 309–322 in Quantitative Methods in Landscape Ecology, M.G. Turner and R.H. Gardner, eds. New York: Springer-Verlag.

Sedjo, R.A. 1998. Forest Service Vision: Or, Does the Forest Service Have a Future? Discussion Paper 99-03. Washington, DC: Resources for the Future.

Sengupta, R., D.A. Bennett, and S.E. Kraft. 2000. Evaluating the impact of policy-induced land use practices on non-point source pollution using a spatial decision support system. Water International 25:437–445.

Tauxe, R.V. 1997. Emerging foodborne diseases: An evolving public health challenge. Emerging Infectious Diseases 3(4):425–434.

US Congress. 1980. P.L. (Public Law) 96-517. Government Patent Policy Act of 1980.

US Congress. 2001. StarLink Corn Controversy: Background. CRS Report for Congress No. RS20732. Washington, DC: Congressional Research Service.

US Congress. 2002. Farm Security and Rural Investment Act of 2002. H.R. 2646.

USDA (US Department of Agriculture). 1994. The Revised ERS County Typology. Rural Research Report No. 89 by P. Cook and K. Mizer. Washington, DC: Economic Research Service, US Department of Agriculture.

USDA (US Department of Agriculture). 1999. Agricultural Resource Management Survey (ARMS). Washington, DC: Economic Research Service, US Department of Agriculture.

USDA (US Department of Agriculture). 2000. Agricultural Resources and Environmental Indicators. Washington, DC: Economic Research Service, US Department of Agriculture. Available online at *http://www.ers.usda.gov/Emphases/Harmony/issues/arei2000/*.

USDA (US Department of Agriculture). 2001a. Food and Agriculture Policy: Taking Stock for the New Century. Washington, DC: US Department of Agriculture.

USDA (US Department of Agriculture). 2001b. Food Spending in American Households, 1997–98, Statistical Bulletin No. 972 by N. Blisard. Washington, DC: Economic Research Service, US Department of Agriculture.

USDA (US Department of Agriculture). 2001c. Structural and Financial Characteristics of US Farms: 2001 Family Farm Report. R.A. Hoppe, ed. Agriculture Information Bulletin No. 768. Washington, DC: Resource Economics Division, Economic Research Service, US Department of Agriculture.

USDA (US Department of Agriculture). 2001d. United States Farm and Farm-Related Employment, 1997. Washington, DC: Economic Research Service, US Department of Agriculture. Available online at *http://www.ers.usda.gov/Data/FarmandRelatedEmployment/ViewData.asp?GeoAreaPick=STAUS_United+States&YearPick=1997&B1=Submit.*

USDA (US Department of Agriculture). 2002a. Food CPI, Prices, and Expenditures. Washington, DC: Economic Research Service, US Department of Agriculture. Available online at *http://www.ers.usda.gov/Briefing/CPIFoodAndExpenditures/Data/table7.htm.*

USDA (US Department of Agriculture). 2002b. Historical Track Records, April 2002: United States Crop Production. Washington, DC: US Department of Agriculture, National Agricultural Statistics Service. Available online at *http://www.usda.gov/nass/pubs/trackrec/trackrec2002.pdf.*

USDC (US Department of Commerce). 2002. US Census 2000. Summary File 3 (SF3) Table P5. Washington, DC: US Bureau of the Census, US Department of Commerce. Available online at *http://factfinder.census.gov/servlet/DatasetTableListServlet?_ds_name=DEC_2000_SF3_U&_type=table&_lang=en&_program=DEC&_ts=51610135218.*

USDHHS (US Department of Health and Human Services). 1998. Preventing Emerging Infectious Diseases: A Strategy for the 21st Century. October. Atlanta, GA: Centers for Disease Control and Prevention.

Unnevehr, L.J., and T. Roberts. 2002. Food safety incentives in a changing world food system. Journal of Food Control 13:73–76.

# 2

# The US Department of Agriculture Research, Education, and Economics Mission Area

The Research, Education, and Economics (REE) mission area is one of seven in the US Department of Agriculture (USDA). It was established to synthesize and circulate knowledge that addresses a broad array of agricultural research subjects. According to the most recent strategic plan, the mission of REE programs is to create a safe, sustainable, competitive US food and fiber system and strong, healthy communities, families, and youth through integrated research, analysis, and education (USDA, 2002).

In the context of the federal research establishment, REE falls at the interface of research areas linked to a variety of other federal agencies, including food, health, nutrition, and disease (Centers for Disease Control and Prevention, Food and Drug Administration, and the National Institutes of Health), environment and natural resources (the Forest Service, the Environmental Protection Agency, the Department of the Interior), energy (the Department of Energy [DOE]), commerce (the Department of Commerce and its National Oceanic and Atmospheric Administration), and basic research (the National Science Foundation, DOE). REE's comparative advantage among federal research institutions lies in its historical strengths in understanding the agricultural and food system. The National Institutes of Health (NIH) may work on specific nutritional components, but USDA is better positioned to assess diets holistically. For example, NIH might work on how lycopene prevents cancer, but USDA can assess how much lycopene is consumed by different subpopulations and how dietary trends or substitutions would affect that intake. Similarly, the interface between agriculture and natural-resource use argues for USDA research to address the environmental impacts of agriculture. No other public agency has the resources, infrastructure, or mandate to support research focusing on the interface between agriculture and the environment. And this is where private-sector research is highly unlikely to fill the void.

## REORGANIZATION OF THE US DEPARTMENT OF AGRICULTURE

In 1994, the Federal Crop Insurance Reform and Department of Agriculture Reorganization Act (US Congress, 1994b) authorized the secretary of agriculture to appoint an undersecretary for research, education, and economics. The explicit purpose of the legislation was to achieve greater efficiency, effectiveness, and economies in the organization and management of the programs and activities carried out by the department (US Congress, 1994b). The legislation delegated all functions and duties related to research, education, and economics to the new mission area, including those previously conducted in other agencies or mission areas. The REE agencies are required to work with USDA action and regulatory agencies: the Agricultural Marketing Service; the Animal and Plant Health Inspection Service; the Foreign Agricultural Service; the Food and Nutrition Service; the Farm Service Agency; the Food Safety and Inspection Service; the Grain Inspection, Packers, and Stockyards Administration; the Natural Resources Conservation Service; the Rural Business-Cooperative Service; and the Rural Utilities Service.

The 1994 reorganization brought four agencies into the REE mission area: the Agricultural Research Service (ARS), the Cooperative State Research, Education, and Extension Service (CSREES), the Economic Research Service (ERS), and the National Agricultural Statistics Service (NASS) (Figure 2-1). Before the consolidation of agencies, ERS and NASS were in the Office of Science and Education. The Forest Service, once a part of the Office of Science Education, was transferred to the Natural Resources and the Environment Mission Area. During this transition, the CSREES was created when the Cooperative State Research Service and the Extension Service merged.

With the passage of the 1996 farm bill (US Congress, 1996), three USDA advisory boards—the Agricultural Science and Technology Review Board, the Joint Council on Food and Agricultural Sciences, and the National Agricultural Research and Extension Users Advisory Board—were consolidated, forming the National Agricultural Research, Extension, Education, and Economics Advisory Board. The new board was mandated to review the REE strategic plan and its implementation and was charged with assisting the secretary of agriculture in creating a task force that would review USDA research. The board also was authorized to counsel the undersecretary with respect to the oversight of REE programmatic content (USDA, 1996, 2002).

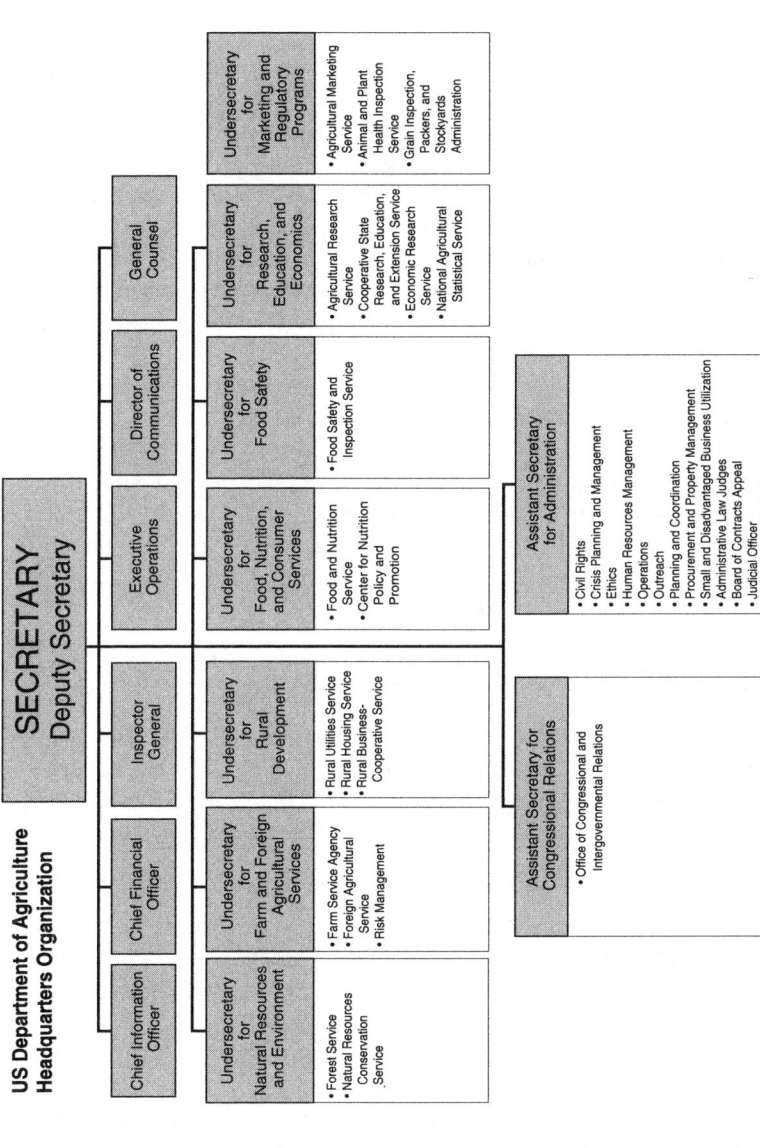

**FIGURE 2-1** US Department of Agriculture Headquarters Organization. Source: USDA, available online at *http://www.usda.gov/agencies/agchart.htm*.

## FUNCTIONS AND STRATEGIC OBJECTIVES OF THE RESEARCH, EDUCATION, AND ECONOMICS MISSION AREA

According to the REE strategic plan (USDA, 2002), the agencies conducting the REE mission area programs perform five primary functions:

1. "Create basic research knowledge at the frontiers of the biological, physical, and social sciences,
2. "Produce, apply, and adopt applied research-based knowledge in innovative ways to address problems and issues,
3. "Produce developmental research results and promote the commercialization and transfer of technologies and practices to potential users in a timely, cost-effective manner,
4. "Provide leadership in the delivery of research-based knowledge through Extension, outreach, and information to strengthen the capacity of public and private decisionmakers, and
5. "Strengthen capacity of institutions of higher education to develop the skills of the Nation's workforce."

The REE mission area's strategic plan delineates five desired outcomes and associated strategic objectives (see Box 2-1; USDA, 2002). The REE strategic plan also assigns resources to the five program objectives in the following distribution: 25% each to the first and second, 16% to the third and fourth, and 18% to the fifth (based on FY 1996 appropriations).

## RESEARCH, EDUCATION, AND ECONOMICS AGENCIES

The four agencies implementing the REE mission, under the leadership of the REE undersecretary, are ARS, CSREES, ERS, and NASS. The history and mission of each agency are described below.

### Agricultural Research Service

ARS, the principal inhouse research agency of USDA, was authorized by the Organic Act of 1862 (US Congress, 1862). The act called upon the then-commissioner of agriculture to "acquire and preserve in his Department all information he can obtain by means of books and correspondence, and by practical and scientific experiments. . . ." With the onset of World War II, the department reconfigured its research units to form the Agricultural Research Administration (ARA). In 1953, ARA underwent a realignment to become what is known today as ARS (USDA, 1999, 2000a).

The ARS mission is "to conduct research to develop and transfer solutions to agricultural problems of high national priority and provide information access and dissemination to: ensure high-quality, safe food, and other agricultural products; assess the nutritional needs of Americans; sustain a competitive agricultural economy; enhance the natural resource base and the environment; and provide economic opportunities for rural citizens, communities, and society as a whole" (USDA, 1999, 2000a). Agency research focuses on achieving the five broad goals identified in the REE Strategic Plan (see Box 2-1) and is organized into 22 national programs (see Box 2-2). ARS is the largest REE agency in overall program staff and budget. According to USDA's Office of Budget and Program Analysis, ARS's budget accounted for half the appropriation for the entire REE mission area in FY 2002.

ARS also is responsible for the administration of the National Agricultural Library (NAL) and the National Arboretum. Brought under the auspices of ARS in 1994, NAL is the world's largest agricultural library. Mandated by Congress to carry out research and provide educational information, the National Arboretum conducts research at four sites in the eastern United States (USDA, 1999, 2000a).

**Cooperative State Research, Education, and Extension Service**

CSREES was formed in 1994 through the USDA reorganization act (US Congress, 1994b). CSREES research and education programs are conducted in partnership with US universities (USDA, 2000b). CSREES's mission is "to achieve significant and equitable improvements in domestic and global economic, environmental, and social conditions by advancing creative and integrated research, education, and extension programs in food, agricultural, environmental and related life and social sciences in partnership with both the public and private sectors" (USDA, 2000b).

The research and education activities of CSREES were authorized under the Hatch Act of 1887, as amended (US Congress, 1887); the Cooperative Forestry Research Act of 1962, as amended (US Congress, 1962); Public Law 89-106, Section (2), as amended (US Congress, 1965); the National Agricultural Research, Extension, and Teaching Policy Act of 1977, as amended (US Congress, 1977b); and the Equity in Educational Land-Grant Status Act of 1994 (US Congress, 1994a). Under those authorities, CSREES assists research and education programs in a research community at state institutions, including the state agricultural experiment stations, schools of forestry, 1890 land-grant institutions and Tuskegee University, colleges of veterinary medicine, and other eligible institutions. It supports competitively awarded and formula-based research programs and regularly scheduled program reviews of the land-grant university agricultural and related sciences.

CSREES is charged with implementing USDA's higher-education mission in the food and agricultural sciences. Its assistance in providing educational opportunities was traditionally less formal until USDA enacted the National Needs Graduate Fellowships Grant Program in 1984. Later initiatives, such as

## BOX 2-1 REE Desired Outcomes and Strategic Objectives

**An Agricultural System That Is Highly Competitive in the Global Economy**
1. Facilitate informed decisions by agricultural producers, policy officials, and other decisionmakers by developing and sharing knowledge promoting agricultural production and marketing.
2. Expand the knowledge base leading to improvements in productivity and marketability, development of new and enhanced commercial products, and expansion of foreign and domestic market opportunities.
3. Ensure the long-term economic viability and sustainability of production agriculture as it makes the transition from federal subsidies to world market orientation.
4. Strengthen and coordinate the capabilities of the REE agencies to enable joint action and rapid response to emerging issues and problems in a global context.

**A Safe and Secure Food and Fiber System**
1. Reduce the impact of threats to agricultural production by expanding the knowledge base needed to rapidly and effectively manage pests, disease, and natural disasters.
2. Improve food safety by developing efficient and reliable monitoring and testing methods to support Hazard Analysis and Critical Control Point (HACCP) and other innovative approaches to food handling and processing.
3. Promote effective and efficient implementation of food-safety policies through research on the economic and socioeconomic impacts of these policies on food production, food processing, and the consumer sectors.
4. Conduct research and adaptive studies to develop integrated production management systems that incorporate HACCP or ISO 9000 standards and ensure meeting sanitary and phytosanitary requirements of the global market.

**Healthy, Well-Nourished Children, Youth, and Families**
1. Reduce disease prevalence and enhance quality of life by defining the relationship between diet, inheritance, and lifestyle and the risk of chronic diseases, acute infections, and immune disorders.

2. Improve the scientific basis for more effective federal food assistance programs by better defining nutrient requirements and monitoring food and nutrient consumption; identifying socioeconomic, cultural, and environmental forces that influence eating habits; analyzing the effect of nutrition information on food choices and diets; and analyzing alternative policies and programs to assist less advantaged citizens in achieving a healthy diet.
3. Generate a more nutritious food supply by conducting research to modify the health-promoting properties of plant and animal foods.
4. Enhance public understanding of diet's role in lifelong health through nutrition education.

**Greater Harmony Between Agriculture and the Environment**
1. Promote sustainable agricultural production and enhance environmental quality by enabling producers to use cost-effective, environmentally friendly production practices and systems.
2. Ensure that policy-makers and program managers have timely, objective data and analysis on the efficacy, efficiency, and equity aspects of alternative agricultural, resource, and environmental policies and programs.

**Enhanced Economic Opportunity and Quality of Life for Citizens and Communities**
1. Promote the effectiveness of rural policies and programs by (a) enhancing understanding of the conditions that promote economic opportunities and (b) identifying rural needs.
2. Promote new businesses and growth in existing businesses, including farms and ranches, by transferring knowledge and technologies developed by or in partnership with REE agencies to private-sector entrepreneurs.
3. Enhance economic opportunity and well-being including the well-being of at-risk children, youth, and families through promoting the use of knowledge by public and private decisionmakers.

Source: REE 2002 Strategic Plan (USDA, 2002).

> **BOX 2-2 The 22 National Programs of ARS**
>
> Food Animal Production
> Animal Health
> Arthropod Pests of Animals and Humans
> Animal Well-Being and Stress Control Systems
> Aquaculture
> Human Nutrition
> Food Safety (animal and plant products)
> Water Quality and Management
> Soil Resource Management
> Air Quality
> Global Change
> Rangeland, Pasture, and Forages
> Manure and Byproduct Utilization
> Integrated Agricultural Systems
> Plant, Microbial, and Insect Genetic Resources, Genomics, and Genetic Improvement
> Plant Biological and Molecular Processes
> Plant Diseases
> Crop Protection and Quarantine
> Crop Production
> Quality and Utilization of Agricultural Products
> Bioenergy and Energy Alternatives
> Methyl Bromide Alternatives
>
> Source: ARS Web site. Available online at *http://www.nps.ars.usda.gov/.*

the Hispanic Serving Institutions Education Grants Program and the Multicultural Scholars program, have been implemented to strengthen the quality of education programs (USDA, 2000b).

The Cooperative Extension System is a national education network of partners from CSREES, land-grant university cooperative extension services, and cooperative extension services in the 3,150 counties of the United States. Cooperative extension work is authorized by the Smith-Lever Act of 1914, as amended (US Congress, 1914), and by Title XIV of the Food and Agriculture Act of 1977 (US Congress, 1977a).

Another important function of CSREES is the collection and integration of national and state data to monitor accomplishments in research, extension, and education, including the Current Research Information System and the Food and

Agricultural Education Information System. Efforts are under way to establish a Research, Education, and Economics Information System to facilitate evaluations of REE activities (USDA, 2000b).

According to USDA's Office of Budget and Program Analysis, the CSREES budget authorizations accounted for 42% of the total REE budget authorization in FY 2002.

## Economic Research Service

ERS is the primary intramural social-science unit in USDA. It was created in 1961 out of the now-defunct Bureau of Agricultural Economics, which was authorized principally by the Agricultural Marketing Act of 1946 (US Congress, 1946). The ERS mission is "to inform and enhance public and private decision-making on economic and policy issues related to agriculture, food, the environment, and rural development" (USDA, 2000c, 2001a), and the agency shares the five goals of the REE mission area, as described above.

According to its FY 2002 performance plan, ERS's goals include research and development of economic and statistical indicators on a broad array of topics: global marketing conditions, trade restrictions, agribusiness concentration, farm and retail food prices, food assistance, foodborne illnesses, food labeling, nutrition, agrichemical use, livestock-waste management, conservation, sustainability, genetic diversity, biotechnology, technology transfer, rural infrastructure, and agricultural labor (USDA, 2000c, 2001a).

According to USDA's Office of Budget and Program Analysis, the ERS budget authorizations accounted for 2.8% of the total REE budget authorization in FY 2002.

## National Agricultural Statistics Service

NASS derives its mandate from the Organic Act of 1862; agricultural supply information was one of the purposes of USDA. The Agricultural Marketing Act of 1946 authorizes NASS's responsibilities (US Congress, 1946). The Census of Agriculture Act of 1997 (US Congress, 1997) reassigned the national census of agriculture from the Department of Commerce to USDA, and NASS was charged with administering the census every 5 years. The agency has about 1,350 federal and state employees, about two-thirds of whom are field operatives (USDA, 2000d, 2001b) and use the state statistical office in 46 states, which are cooperatively funded through land-grant universities or state departments of agriculture. Other state institutions are also eligible for NASS services related to survey administration, and the agency undertakes some international activities (USDA, 2000d, 2001b).

NASS states that its mission is to "provide timely, accurate, and useful statistics in service to US agriculture" (USDA, 2000d, 2001b). Its clients are drawn from almost every aspect of the farming and agribusiness communities. Clients typically provide NASS with responses that are used to generate data concerning such subjects as crop and livestock counts. The Agricultural Statistics Board of NASS uses the information to produce several hundred reports per year (USDA, 2000d, 2001b).

In addition to the farming and agribusiness members that it serves, NASS's statistical information and reports are used in the decision-making processes of the White House, Congress, other USDA programs, and other federal agencies. Among the specific NASS activities that contribute to those processes are the supply of data used for the world agricultural supply and demand estimates, the Export Enhancement Program, the Conservation Reserve Program, the determination of milk prices, grazing fees (on publicly owned land), and USDA's crop-insurance program (USDA, 2000d, 2001b).

According to USDA's Office of Budget and Program Analysis, the NASS budget authorizations accounted for 4.6% of the total REE budget authorization in FY 2002.

## SUMMARY

This chapter has provided background information on the 1994 USDA reorganization that brought about the REE mission area. The functions and strategic objectives of the mission area, the history and mission of each agency, and the proportion of the total REE budget accounted for by each agency were also considered.

## REFERENCES

US Congress. 1862. 7 USC. 2201. Organic Act of 1862.
US Congress. 1887. 7 USC. 361a et seq., 24 Stat. 440. Hatch Act of 1887.
US Congress. 1914. 7 USC. 341 et seq., 38 Stat. 372. Smith-Lever Act of 1914.
US Congress. 1946. 7 USC. 1621–1627. Agricultural Marketing Act of 1946.
US Congress. 1962. Cooperative Forestry Research Act of 1962, as amended (16 USC. 585a–7).
US Congress. 1965. P.L. (Public Law) 89-106, Section (2), as amended (7 USC. 450i).
US Congress. 1977a. P.L. (Public Law) 95-113. Food and Agriculture Act of 1977 (7 USC. 2281. 91 Stat. 1041).
US Congress. 1977b. P.L. (Public Law) 95-113. National Agricultural Research, Extension, and Teaching Policy Act of 1977.
US Congress. 1994a. P.L. (Public Law) 103-382. Equity in Educational Land-Grant Status Act of 1994.
US Congress. 1994b. P.L. (Public Law) 103-354. Federal Crop Insurance Reform and Department of Agriculture Reorganization Act of 1994.
US Congress. 1996. P.L. (Public Law) 104-127. Federal Agriculture Improvement and Reform (FAIR) Act of 1996.
US Congress. 1997. P.L. (Public Law) 105-113. Census of Agriculture Act of 1997 (7 USC. 2204 g).

USDA (US Department of Agriculture). 1996. Charter of the National Agricultural Research, Extension, Education, and Economics Advisory Board. Washington, DC: Government Printing Office. Available online at *http://www.reeusda.gov/ree/advisory/charter.html*.

USDA (US Department of Agriculture). 1999. Agricultural Research Service Strategic Plan: Working Document 1997–2002. Washington, DC: Agricultural Research Service, US Department of Agriculture. Available online at *http://www.nps.ars.usda.gov/mgmt/stratpln/1999/background.cfm*.

USDA (US Department of Agriculture). 2000a. Agricultural Research Service FY 2000 and 2001 Annual Performance Plans. Washington, DC: Agricultural Research Service, US Department of Agriculture.

USDA (US Department of Agriculture). 2000b. Cooperative State Research, Education, and Extension Service FY 2000 and 2001 Annual Performance Plan. Washington, DC: Cooperative State Research, Education, and Extension Service, US Department of Agriculture.

USDA (US Department of Agriculture). 2000c. Economic Research Service Strategic Plan, 2000–2005. October 11, 2000. Washington, DC: Economic Research Service, US Department of Agriculture. Available online at *http://www.ers.usda.gov/AboutERS/ersstrategicplan.pdf*.

USDA (US Department of Agriculture). 2000d. National Agricultural Statistics Service. GPRA Strategic Plan, Washington, DC: National Agricultural Statistics Service, US Department of Agriculture. Available online at *http://www.usda.gov/nass/nassinfo/strat-2005.pdf*.

USDA (US Department of Agriculture). 2001a. Economic Research Service FY 2002 Annual Performance Plan and Revised Plan for FY 2001 (July). Washington, DC: Economic Research Service, US Department of Agriculture. Available online at *http://www.ers.usda.gov/AboutERS/ ersperformance_plan.pdf*.

USDA (US Department of Agriculture). 2001b. National Agricultural Statistics Service FY 2002 and Revised FY 2001 Annual Performance Plans. Washington, DC: Economic Research Service, US Department of Agriculture. Available online at *http://www.usda.gov/nass/nassinfo/nass-app-02-01.pdf*.

USDA (US Department of Agriculture). 2002. Research, Education, and Economics Strategic Plan. Washington, DC: US Department of Agriculture. Available online at *http://www.reeusda.gov/ ree/ree2.htm*.

# 3

# Research Frontiers

The demands for research to support continued productivity gains, more and varied products, better human health, enhanced biosecurity, animal welfare, environmental benefits, and the vitality of rural communities are growing. At the same time, scientific advancement, innovation, and technologic development in a variety of fields, from molecular biology to ecosystem dynamics, offer new opportunities for research to meet the demands. The US Department of Agriculture's (USDA) Research, Education, and Economics (REE) mission area is uniquely positioned to carry out research in these frontier areas that will serve important public goals.

Agricultural research can address issues arising from five major phenomena: globalization; emergence of pathogens; links between diet, health promotion, and disease prevention; the relationship between agriculture and the environment; and changes in rural communities. This chapter highlights research directions related to each of those challenges that

- Provide broad benefits for agriculture, the environment, and US citizens, families, and communities.
- Anticipate the future and capture the unique opportunities of our time.
- Enhance the global competitiveness of the US food and agricultural system.
- Push the REE research agenda to be more consumer-driven rather than production-driven.

## GLOBALIZATION

Few recent economic changes equal those brought about by the globalization of the US economy in the last quarter of the 20th century. Now, in addition to managing their highly productive resource base, US agriculturalists must respond to changing consumer demands for products and services and must manage technology, capital, and labor in globally integrated markets. Even with slowing worldwide population growth, demand for livestock products will rise dramatically with income growth in less-developed countries and lead to new market opportunities and new global challenges to agricultural systems (Delgado et al., 2001). To be competitive in this global economy, US agriculture will need to continue its technologic leadership and long-term productivity gains. That will require new and more sophisticated technologies and systems for managing information. Advances in information technology and in genomic sciences create new possibilities for research to aid agriculture in delivering higher-quality products and services. But as the global nature of potential risks posed by new technology is better understood, there is also a need for more sophisticated evaluation of such risks. Thus, globalization creates the demand for greater understanding of how global forces affect US agriculture, continued improvements in agricultural productivity, and better ex ante evaluation of risks posed by new technology.

### Evaluate the Implications of Globalization for US Agriculture and Agricultural-Research Priorities

The worldwide trend for countries to export and import a growing share of goods, services, factors of production,[1] and intellectual property will have important effects on national economies, societies, and the environment. Research is needed to provide a sound, scientific basis of policies and programs that address those effects in the United States. Such research must be integrative and examine the full effects of globalization and the environmental, social, and economic trade-offs that policy-makers will face. One of the principal issues that research should address is the relative benefits and costs of investing in different kinds of research, including research that yields societal and environmental benefits. A second issue is the challenge of removing policy distortions that bias incentives in world agriculture. A third issue is the changing international balance of supply and demand, including the continuing lack of food security[2] in many nations.

---

[1] Factors of production are the resources available for producing goods, and typically include land, labor, and capital. They may also include other natural resources, entrepreneurial ability, and human capital.

[2] According to the Life Sciences Research Office, Federation of American Societies for Experimental Biology, food security exists when all people at all times have access to enough food for an active

These three issues are linked both globally and domestically. Although such research is currently undertaken by REE agencies, the scope of these issues will require REE agencies to break from convention and undertake research that is broader and more multidisciplinary and that involves collaborative partnerships with diverse institutions and agencies in the United States and internationally.

A related area of research is better understanding of how worldwide changes in intellectual property rights policy alter the public research agenda. Changes in technology, in legal rulings, and in international agreements have increased the return on investment from privately funded agricultural and food research and the international spillovers from research investments (Parker et al., 2001; Reilly and Schimmelpfenning, 2000). Partnerships, joint ventures, and other alliances between public and private institutions are becoming more common in agricultural research. Such partnerships increase funding for some kinds of research and improve the prospects for commercialization and use of new technologies, but at the same time they raise concerns about whether private-sector interests are playing too great a role in setting research priorities (Knudson, 2001; also see Chapter 5). Although such concerns are not peculiar to agriculture (e.g., Feller et al., 2002; Heller and Eisenberg, 1998), the pace of change in agricultural research institutions and in biotechnology raises many unresolved issues (Smith, K.R., et al., 1999). For example, will so-called interlocking patents on components of new technologies or knowledge prevent applications to new discoveries when they are owned by different parties (Smith, K.R., et al., 1999)? What role could the public sector play in bringing these parties together for discoveries with broad public benefit? Research is needed to understand better which new strategies for research funding, public–private collaboration, and technology transfer will yield the highest return on the public research investment.

### Improve Agricultural Productivity and Product Quality While Optimizing Resource Use

Conventional approaches to genetic improvement have successfully enhanced the productivity, disease resistance and pest resistance, nutritive quality, and safety of plants and animals. Further improvements are now possible through genomics- and proteomics-based technologies. Although commercial investment in biotechnology is high, REE should continue to have a key role in research that is unlikely to be well supported by the private sector.

For example, REE must lead the preservation of the nation's agricultural

---

and healthy life. This includes at a minimum (1) the ready availability of nutritionally adequate and safe foods and (2) the assured ability to acquire acceptable foods in a socially acceptable way (for example, without resorting to emergency food supplies, scavenging, stealing, or other coping strategies) (FASEB LSRO, 1990).

genetic resources. The public sector also must invest in research to improve the efficacy and specificity of gene-transfer technology. Important research includes developing techniques for modifying plant and animal genomes, building models and systems analyses that integrate basic knowledge about plants and animals into gene selection, and synthesizing research findings on gene mapping and the expression of proteins associated with quantitative traits (proteomics). Current understanding of physiologic mechanisms and metabolic pathways does not provide sufficient precision for targeting genetic manipulations. Given the high cost of genetic manipulations, especially in animals, greater precision and predictability are essential. Collaboration among experimentalists and modelers will be essential to develop quantitative and dynamic models of interactions in physiologic and metabolic systems; this will enable scientists to make specific improvements and to understand the implications for the entire organism better.

Finally, the application of genomics-based approaches to environmental issues is unlikely to have high commercial priority and should fall in the public-sector portfolio. Advances in agricultural genomics resulting from research in the above subjects will create new information resources and needs and consequently enlarge the use of bioinformatics in agriculture for acquiring, processing, storing, distributing, analyzing, and interpreting biologic information.

Precision agriculture is another frontier technology that could substantially improve productivity while providing environmental benefits. This spatially explicit approach to crop management involves tracking production and tailoring inputs to meet the specific needs of subacre areas in individual fields. Recent advances in the technologies that underlie precision agriculture have outstripped their practical application. We need workable decision-support tools that will enable farmers to adjust the timing and amounts of seed, fertilizer, water, and pesticides to optimize production while minimizing waste and environmental effects. Close collaboration among experimental scientists, statisticians, economists, engineers, and systems analysts will be essential for integrating experimental research into decision-support systems and underlying models for crop, animal, and environmental systems.

The scientific underpinnings of farming approaches that seek to minimize agricultural inputs and adverse environmental effects—broadly captured by the terms *sustainable*, *alternative*, and *organic*—have burgeoned in recent decades (e.g., Robertson and Harwood, 2001). Despite rapidly expanding consumer demand for organic or low-input agricultural products, funding of related research by the agricultural-technology sector has been chronically low because few discoveries can be commercialized. Consequently, REE must play a critical role in supporting both fundamental research on the functioning of agroecosystems and applied research on methods of enhancing production by modifying or augmenting agroecosystem processes. Other important research will include assessments of economic competitiveness and barriers to user adoption of such farming practices.

### Evaluate the Economic, Social, Health, and Environmental Effects of Agricultural Technologies and Practices

Understanding the full potential effects—social, economic, health, environmental, and ethical—of new technologies and practices, including their global effects, is crucial to sound research choices and to technology transfer. New technologies often have enormous promise to enhance people's lives. However, they also raise important questions about environmental and health risks, the distribution of benefits and risks, and public values and ethics. Exploring such questions early in the R&D process will focus investment in technology development on efforts most likely to generate the greatest public benefits.

The production of genetically modified food, for example, has raised new issues related to the appropriate level of health and environmental review, product labeling, and public communication. Public debate has highlighted differences in perceptions and values among segments of society and among scientists who have different expertise. Other emerging technologies and practices will raise similar issues. Recent and current examples include the use of recombinant bovine somatotropin in dairy cattle, development of antibiotic resistance from use of antimicrobials in the livestock and dairy industries, the causes of and solutions to coastal hypoxia, and the availability and uses of human genetic information.

Optimizing the benefits of new agricultural technologies and practices will require research on risk assessment and communication, applied ethics, public values, and negotiated decision-making processes. Some efforts, such as those to assess the ecologic effects of new technologies and practices on near and distant ecosystems, will require research to develop more effective analytic frameworks and methods.

Publicly supported research on new technologies must be coupled with public education that demystifies scientific and technical information for the general public and provides balanced information about benefits and risks. Public-education efforts should be coupled with social-science research and discussion to ensure that information about public understanding and values is incorporated into the initial stages of new technology R&D. REE is uniquely positioned to provide leadership in this respect because of its dual responsibilities for research and education.

## EMERGING PATHOGENS AND OTHER HAZARDS IN THE FOOD-SUPPLY CHAIN

Advances in the science of public health, changes in how consumers obtain and prepare food, and increases in international trade in food products and animals all increase the profile of food safety and animal and plant health (Unnevehr and Roberts, 2002). Preharvest and postharvest foodborne pathogens—such as

*Campylobacter jejuni, Salmonella enteritidis, Listeria monocytogenes,* and *E. coli* O157:H7—continue to emerge and to pose threats to human health (Hughes, 2001; Todd, 2001; Unnevehr and Roberts, 2002). Furthermore, the long-term consequences of many foodborne illnesses are only now being uncovered, such as the link between salmonella infection and rheumatoid arthritis. Understanding and reducing foodborne risks to human, animal, and plant health will require new research that will ultimately support both private and public efforts to eliminate hazards. New scientific tools, such as genetic "fingerprinting" of microbial pathogens and rapid detection methods, provide new opportunities for epidemiology and risk assessment. The threat of bioterrorism lends urgency to those research needs.

## Reduce the Risks of Bioterrorism

The risk of a terrorist attack on the United States that targets the food or water supply is a critical national concern (Frist, 2002). Several agencies with different and complementary expertise are collaborating to reduce the threat and to increase our capability to minimize the loss of life and other consequences if such a disaster occurs.

REE is already a key contributor to collaborative federal efforts against bioterrorism, and the demand for further contributions will increase in the decades ahead. The growing international trade in food products and ingredients will multiply the number of possible points of introduction of harmful agents into nonprocessed and processed foods, and the virulence of emerging and potential pathogens heightens the risk. But REE's ability to provide the research needed to avert a biologic attack via the food or water supply has declined in recent years because of reduced funding. There is an unprecedented need for scientists with appropriate training and for upgraded facilities to conduct biohazard research. Within REE are laboratories that would be high-priority candidates for improved security.

## Improve Microbiologic Food Safety

Serious gaps persist in the nation's ability to rapidly and effectively manage known and emerging preharvest and postharvest pathogens, that is, to detect, trace the origins of, and eliminate pathogens in the farm-to-table food chain. Although recent research has improved food safety and the US food supply is one of the safest in the world, the system's growing complexity and dynamism continue to generate needs for information (Kuzminski, 1994). For example, current food-consumption trends toward more fresh, uncooked, fast, and imported foods raise questions about the sources of and solutions to food contamination (Hughes, 2001; Todd, 2001; Unnevehr and Roberts, 2002). At the same time, improved scientific understanding of pathogen evolution and virulence from genomics research has opened important new research avenues related to the identification and origins of pathogens. Research on the epidemiology and public-health consequences of microbial pathogens must be integrated with research on control and monitoring of pathogens. Multidisciplinary research for risk assessment, risk management, and risk communication has the potential to make a major contribution to the safety of the US food supply. Such research must be dynamic and evolving if it is to "anticipate future microbial hazards and construct barriers to disease" (IFT, 2002). Timely application of new discoveries will assist the USDA action agencies in addressing their own emerging needs through applied research.

**Understand and Minimize the Hazards of Food Allergens and Toxicants**

Food allergens[3] and toxicants[4] and their mechanisms of action are poorly understood, and this hampers the development of prevention strategies and therapies (FDA, 1992; NRC, 2000b). Improved knowledge, including adequate methods for screening novel allergens or toxicants, is increasingly urgent in light of the concern that transgenic or conventional breeding technologies may create unexpected allergenic or toxic properties in food through pleiotropic processes (NRC, 2000b). Moreover, it is uncertain whether transgenic techniques are more likely than conventional plant-breeding techniques to increase the risks related to allergens, toxicants, or other unintended consequences (NRC, 2000b). Two examples of unexpected allergenic or toxic properties of transgenic technologies are the transfer of potential allergenicity from a Brazil nut gene introduced into soybean to enhance its nutritional content (Nordlee et al., 1996) and the *Bacillus thuringiensis* Cry 9C protein, which does not degrade rapidly in gastric fluids and raised concerns of potential allergenicity when it was inadvertently introduced into the human food supply (USDHHS, 2001; USEPA, 1998).

Insofar as research related to the creation of transgenic crops has greatly outpaced research related to pleiotropic and other unintended consequences, there is strong public and scientific interest in creating a government-sponsored program to explore questions about food allergens and toxicants that are unlikely to be pursued by the private sector. An aggressive federally funded program would speed necessary basic research, for example, developing an animal model of food allergenicity in humans. Once these questions are resolved, it may be possible to identify the mechanisms by which some proteins cause allergies or toxic effects and to develop innovative mechanisms to reduce the hazard associated with these proteins. The mechanisms might include developing biotechnologic approaches to inactivate allergenic or toxic substances in foods.

---

[3]Food allergens may include peanut, shellfish, milk, and eggs.

[4]A food toxicant is a naturally occurring chemical (a chemical produced by a plant or animal) that is harmful. Glycoalkaloids in potatoes and furanocoumarins in celery are examples (NRC, 2000b).

### Improve Understanding and Management of Plant and Animal Diseases

Advances in science offer new opportunities to manage plant and animal health in an increasingly integrated global economy. They include new applications of epidemiology, risk assessment, and risk-management tools to understand risks posed by wildlife or by increased international trade in plants and animals. Enhancing disease resistance of plants and animals through genetic techniques could yield major benefits by reducing processing and production costs and lessening the use of antibiotics in animal production. Basic research on applying biotechnology will be a requisite for such applied research. REE should also support research on other alternatives to antibiotics for promoting growth and preventing livestock disease, such as competitive exclusion and vaccination, to address questions about the human health implications of antibiotic use in livestock and producers' desires for improved management options.

## NUTRITION AND HUMAN HEALTH

Despite food and nutrition assistance programs, hunger and food insecurity persist in the United States. Food-insecurity prevalence was 10.8% across households in the United States during the period 1998–2000, with prevalence ranging from 7.8% to 15.9% of households among the states (Sullivan and Choi, 2002). In addition, prevalence of overweight and obesity[5] among US adults has increased over the last 3 decades and was estimated at 61% in 1999 (USDHHS, 1980, 1988, 1999), and the percentage of overweight children and adolescents has also increased (USDHHS, 1970, 1974). Many chronic diseases are weight-related, including diabetes, cancer, heart disease, stroke, hypertension, gallbladder disease, osteoarthritis, sleep apnea, and asthma. Weight-related behaviors, such as poor diet and lack of physical activity, are linked to these continuing epidemics (Mokdad et al., 2001). To date, the primary US policy response to long-term diet-related conditions, such as obesity and chronic disease, has focused on consumer information (for example, through labeling) and education (for example, through the Expanded Food and Nutrition Education Program), with much less attention given to the community and societal factors that facilitate or inhibit the adoption and maintenance of healthful diets and lifestyles.

There is urgent need for continued REE research to guide and evaluate food and nutrition policies and interventions at multiple levels and settings, including individual, family, school, worksite, retail, marketing, and production. Some of these research priorities are identified in the US Action Plan on Food Security (USDA, 1999b). Many aspects of the links between diet, health, and disease are only now becoming understood. Exciting new possibilities for improving health,

---

[5]Obesity is defined as a body-mass index score of 30 or more. Overweight is defined as a body-mass index score of 25 or more.

for controlling some diseases, and for preventing or postponing the onset of some chronic diseases through diet and for tailoring diets to individual nutritional risks are emerging. Strengthening and expanding such priorities will be one of the most important ways for agricultural research to provide benefits to the general public. Although the development of new food products will be driven by private-sector funding, USDA should expand research to provide a scientific basis for efforts to shift dietary patterns and physical activity in a more healthful direction.

As science evolves and public-health challenges shift, a flexible framework for setting research priorities must be constructed. REE should develop a research strategy that focuses resources on the most prevalent and costly diseases for which research has the greatest potential for improving the health of the American people. The REE effort should be done in collaboration with other public-health agencies, including NIH (see Chapter 5 for additional discussion of collaboration).

### Advance Research on Bioactive Food Components

REE has a tremendous opportunity to evaluate the health effects of biologically active food components that promote health and prevent disease. Bioactive components occur naturally in many foods, especially fruits and vegetables, and include an array of chemical compounds with varied structures, such as carotenoids, flavonoids, plant sterols, omega-3 fatty acids, allyl and diallyl sulfides, indoles, and phenolic acids. There is a need for a scientific understanding of the chemistry, metabolism, and health effects of these food components. There is also a need to assess the concentrations of these components in foods and to incorporate the information into food-composition databases so that dietary intakes may be estimated and tracked. The Agricultural Research Service should continue its work compiling databases on carotenoids, flavonoids, and other bioactive compounds.

### Elucidate Genetic Mechanisms That Affect Human Health and Nutrition

Nutrition-related research on human genetics will provide the foundation for further understanding of the metabolic fate of nutrients and the biochemical functions of food components, including macronutrients, vitamins, minerals, bioactive components, and pharmacologic agents. It also will elucidate how and why people vary in their requirements for and uses of various food components. Such knowledge has important applications to disease prevention and to minimizing exposure to physiologically harmful ingredients in plant and animal products.

The genetic basis of such variation is not well understood. Researchers have identified relatively few of the specific genes that affect the human body's use of various food components. Also unknown are many aspects of how the genes interact with one another or with the environment to produce specific nutritional or disease outcomes. Near-term research priorities include the identification of biomarkers that correlate with gene activity and functional genomics and proteomics research to understand correlations between genotype and phenotype. Such work eventually should make it possible to identify a constellation of phenotypes that signal high disease risk.

Improved understanding of how genes affect individual nutritional status and disease risk could eventually have an important role in shaping public-health policy. For example, a better understanding of how genes affect the body's storage and use of food calories would greatly enhance efforts to develop effective food and nutrition policies for reversing our national epidemic of obesity.

In light of the likely rapid entry of transgenic foods into the marketplace in the coming years and the potential that some of the intended and unintended compositional changes may disproportionately affect genetically susceptible segments of the population (NRC, 2000b), there is some urgency to accelerating the research into the interactions between genes and bioactive compounds in food and dietary supplements. Indeed, there may be merit in coordinating this research in some manner that gives priority to studying the genetic interactions with ingredients that are consumed by the most people or that hold the greatest potential for producing undesirable consequences.

## Improve the Nutrient Content of Foods

Opportunities are expanding to enhance human health through plant and animal products that have improved or enhanced nutrient content. Dietary shifts among consumers toward healthier eating patterns are generating demands for foods of superior health quality (Krauss et al., 1988). Continuation of that trend is expected to reinforce the changes of the last decade that made nutraceuticals and functional foods (foods containing bioactive components) a substantial part of the food industry (Childs, 2001; Van Elswyk et al., 1998).

With those shifts in consumer demand, scientific discoveries have greatly expanded understanding of where and how nutrient enhancement could yield improvements in human health. Through advances in biotechnology, scientists now envision using plants as "nutrient factories" that produce nutritionally fortified foods (Burn and Kishore, 2000; Kleese, 2000;) and using major crops as tools for improving human health (Della Penna, 1999). Similar advances in animal biotechnology and scientific understanding of the controls over animals' physical traits will enable researchers to modify meat composition.

Modification of fats in plant and animal products is a particularly promising research subject because some fat-consumption patterns are thought to affect the risk of cardiovascular disease, cancer, and diabetes in adult humans and to improve health and nervous system development in newborns. Food technology is a direct approach for modifying the fatty acid properties of foods. Processed foods are the primary source of *trans* fatty acids, and processors are already implementing new technology to eliminate these. In addition, today's understanding of the genetic controls over fat structure in plants should make it possible to customize plant lipid biosynthesis to reduce saturates, decrease oxidation potential, eliminate *trans* fatty acids, and increase essential long-chain polyunsaturated fatty acids and antioxidants (Brown at al., 1999).

## Improve Understanding of Food-Consumption Behavior and its Links to Health

National food-consumption surveys and nutritional epidemiology studies have been key components of the current understanding of the relationships between diet, health, and disease. REE has an important continuing role to play in the collection and evaluation of food-consumption data. The USDA Agricultural Research Service and the Department of Health and Human Services National Center for Health Statistics have worked collaboratively to implement the congressionally mandated merger of the National Health and Nutrition Examination Surveys (NHANES) and the Continuing Survey of the Food Intakes of Individuals into a single comprehensive, national food-consumption and health survey (called NHANES). USDA's improved method of obtaining data on food consumption is a critical component of the new merged survey. More-detailed food-consumption data, including data on brand-name processed foods and restaurant foods, will allow better interpretation of the results from NHANES and from other nutrition research studies (such as clinical trials and nutrition-intervention studies).

Subar et al. (in press) and Kipnis et al. (in press) reported that current dietary-assessment methods—24-hour dietary recalls and food-frequency questionnaires—underestimated both protein and energy intake. There is a great need for REE to continue to improve methods of assessing food consumption so that the results will be accurate and provide insight into diet-related health issues, such as obesity, diabetes, some forms of cancer, and other chronic diseases.

Growing public use of dietary supplements (Eisenberg et al., 1998) has created new needs to incorporate related information into REE's food-composition database and food-consumption survey. This information will allow estimations of the extent, level, and types of dietary supplements consumed among various demographic groups and the beliefs and motivations that underlie these behaviors. Data on dietary-supplement consumption may reveal associations between dietary-supplement intake and health measures and safety concerns. There is also a major gap in knowledge of the safety of various ingredients in dietary supplements. The private sector has little incentive to invest in this subject and no regulatory requirement to do so, and it might therefore be appropriate for publicly funded research. This information is also vital for research and public-policy decisions on nutrition-related issues.

Improvements in human nutrition and health will depend on the actions of individuals, households, and food manufacturers. Although private research is extensive, it focuses on enhancing the appeal of food products for targeted consumer markets. A comparable research base does not exist for understanding the reasons for consumer choices related to food selection, exercise patterns, and unhealthy habits, such as alcohol abuse and use of tobacco and recreational drugs. Such a research base would help to answer such questions as, Which public or private sources of information are most credible to various segments of society? What values, beliefs, and environmental factors motivate health- and nutrition-related behaviors? How do motivation, behavior, and environmental factors vary among individuals or population groups? Answers to those questions will be essential for designing effective nutritional policies and programs.

## ENVIRONMENTAL STEWARDSHIP

An important transformation in how the American public views the relationship between agriculture and the environment is under way. Whereas past public policies sought to minimize the harmful effects of agricultural practices on the environment, such as pollution, the policies of today and those of tomorrow will go further toward realizing agriculture's potential to deliver broad environmental benefits, such as clean water, carbon sequestration, and biodiversity conservation. Agricultural research thus must play the dual roles of developing environmentally nonharmful farming practices and advancing new practices for managing land and natural resources that will yield environmental benefits. Both endeavors will be aided by integrating recent conceptual advances from the ecologic and social sciences. In the context of increasing pressures on global land, water, and genetic resources and global environmental change, US-based agricultural research can contribute to delivering global environmental benefits and to informing decision-making on international environmental agreements.

### Reduce Pollution and Conserve Natural Resources

Air and water pollution and its harmful effects on the environment and human health remain important byproducts of many agricultural practices. The sources of pollution include fertilizers and pesticides used to enhance productivity (NRC, 2000a; Smith, V.H., et al., 1999); animal wastes, particularly from animal-feeding operations (CENR, 2000; NRC, 2000a, 2002b); greenhouse gases (IPCC, 2001; Robertson et al., 2000); and soil released by some production methods (e.g., Lal, 1998). Research is needed to understand the off-farm transport of agricultural contaminants and to design more effective strategies for keeping nutrients, chemicals, and soils within the farming system.

National attention has recently focused on invasive species—species spread beyond their natural geographic ranges—as a major threat to agriculture, other industries, public health, and natural ecosystems. More than 300 nonnative weeds have invaded western rangelands (Babbitt, 1998; Di Tomaso, 2000); exotic insects and pathogens (such as the imported red fire ant, the soybean cyst nematode, and the gypsy moth) threaten public health and the productivity of crop and forest ecosystems; and invasive species have contributed to the decline of almost half the imperiled species in the United States (Wilcove et al., 1998). Enormous gaps exist in our ability to predict or mitigate most species invasions, and research needs are great. For example, what makes some ecosystems and certain locations more susceptible to invasions and some species better and more damaging invaders? What are the economic and ecologic trade-offs among different control strategies? Several groups of scientists have summarized key scientific research questions (e.g., Byers et al., 2002; Ewel et al., 1999; Mack et al., 2000; McNeely et al., 2001). Moreover, the National Invasive Species Management Plan, developed by an interagency council convened under Executive Order 13112, has identified high-priority research needs for reducing the economic and environmental impacts of invasive species in the United States (NISC, 2002).

In agricultural areas across the United States, application of fertilizers, manure, and pesticides (primarily herbicides) have degraded the quality of streams and shallow groundwater. Some of the highest concentrations of nitrogen found in recent national water-quality assessments occur in streams and groundwater in agricultural areas (NRC, 2000a; USGS, 1999). Soil erosion from cropland causes substantial losses in topsoil quality and quantity (9 t/ha per year) (Heinz Center, 1999; USDA, 2000a, 2000b) and results in environmental costs estimated to be about $17 billion per year (Pimentel et al., 1995). The use of about 6% of the total US energy budget by the agricultural sector (including use for manufactured inputs, such as fertilizers and pesticides) consumes much imported oil and yields greenhouse gases (Duncan, 2001; Pimentel et al., 2002; USBC, 1999). Research is needed on technologies and policies that will reduce soil erosion, improve water quality, improve the efficiency of cropland irrigation, and increase the energy efficiency of agricultural production.

## Advance Environmentally Sound Alternatives

Reduced pesticide use could result from broader US adoption of pesticide alternatives, including biologic pest control and mechanical crop-management practices, such as crop rotation, cover crops, and tillage and planting techniques that disrupt pest life cycles and promote natural enemies. Adoption of those approaches remains low in the United States in comparison with other nations (NRC, 2000c; Pattersson, 1997; Pimentel et al., 1993). Socioeconomic research is necessary to understand and address the impediments to farmers' adoption of pesticide alternatives.

Reductions in domestic oil supplies have heightened interest in the use of fuels and feedstock derived from corn and nontraditional crops, such as perennial grasses (NRC, 1999). Corn-based ethanol production has received particular attention. Related research on biofuels needs to consider not only biochemical feasibility but also energy efficiency in feedstock production, fermentation, and distillation processes (Pimentel, 2001), and the full range of socioeconomic and environmental costs and benefits of biofuel production and use.

The development of improved crop cultivars and livestock strains through genomics holds the promise of enhancing agriculture's environmental compatibility in numerous ways. Improving plant nutrient use and improving efficiency of nutrient digestion and use in livestock, for example, could help to lower fertilizer needs and keep excess nitrogen and phosphorus out of waterways. Enhanced efficiency of water use by major crops would reduce agricultural water demands. Plants and animals with increased resistance to pests or diseases should require lower rates of pesticide and fungicide application. Related research should focus both on extension of biotechnology to environmental issues and on evaluating related environmental risks, such as the potential spread of novel genes and phenotypes into populations of native microorganisms, plants, and insects. The spread of novel genes to microorganisms in particular poses a largely unknown risk to the ecology of agricultural and natural landscapes.

## Deliver New Environmental Benefits

The increasing public demand for recreational and environmental services from the nation's land and water resources has enormous implications for agriculture and rural economies. One effect has been the clear trend in US agricultural policy, such as the conservation title of the US Farm Security and Rural Investment Act (US Congress, 2002), to reward farmers for delivering environmental benefits beyond food and fiber production. That approach is expected to play an increasingly important role as the nation seeks solutions to some of its most pressing environmental challenges. For example, appropriately managed agricultural lands are expected to be critical in managing the nation's water resources by capturing and filtering rainwater and by sustaining wetlands that dissipate river floodwaters and filter runoff. Carbon sequestration in fields, rangelands, and forests is recognized for its importance in controlling global warming and climate change, and the emergence of carbon-credit markets is creating a new agricultural commodity (CAST, 2000; IPCC, 2000). The long-term conservation of the nation's biologic diversity may depend in part on management of agricultural lands to provide habitats and migratory corridors for native plants and animals.

However, the science underpinning the environmental benefits of agriculture has lagged substantially behind policy advances. Some of the key researchable questions are technical. Which lands should be managed in what manner to deliver what benefits? For example, what configurations of land use will best support biologic diversity in agricultural landscapes? How might the disappearance of farmland due to exurbanization and transportation corridors affect future opportunities? Socioeconomic research also is needed to design policy instruments for encouraging private environmental stewardship. For example, what would be the implications of creating new sources of farm income from such activities as the sale of hunting rights and government payments for sequestering atmospheric carbon?

## Integrate Leading-Edge Environmental Science Concepts and Technologies

Promoting the environmental application of new technologies and of conceptual advances in the biophysical and social sciences should be an integral part of the REE environmental portfolio. Dynamic and innovative research approaches that include those elements could position REE to make important advances against some of the most vexing environmental challenges. For example, the combination of a biophysical understanding of nutrient dynamics and a socioeconomic understanding of policy, legal, and market influences on environmental decision-making will be essential for crafting long-term solutions to hypoxia in the Gulf of Mexico caused by agricultural runoff (Goolsby et al., 2000; NRC, 2000a; Rabalais et al., 1996) and for mitigating greenhouse gases (CAST, 2003).

Environmental research administered by REE requires study on the appropriate geographic and time scales. For example, questions related to water supply and quality often require study at the level of entire watersheds or larger areas rather than the traditional field or farm level of analysis. Similarly, analyses of processes that occur over decades and longer periods may be required to provide solutions that involve natural long-term environmental changes, such as climate variability, biogeochemical processes, and insect, weed, and pathogen dynamics.

The newer geospatial technologies—geographic information systems and the global positioning system—coupled with related analytic and modeling approaches hold great promise for understanding and addressing the underlying links among landscape units in watersheds and regions. The benefits of greater emphasis on spatial analyses would cut across environmental issues, from the spread of invasive species to the design of wetlands and riparian filter strips.

Other new technologies offer additional opportunities. New molecular approaches provide tools for understanding soil microbial communities and managing the complex relationships between soil biology and ecosystem health (Buckley and Schmidt, 2001a, 2001b). Nanoscale technologies could help to reveal the dynamics of biologic processes, such as insect movement, pheromone dynamics, and soil-atmosphere gas fluxes. And advances in information technology, including wireless field monitors and environmental-database design, will help to understand environmental variability.

## QUALITY OF LIFE IN RURAL COMMUNITIES

The quality of life in rural communities is deteriorating in many regions with shifting populations, inadequate workforce competence, and weak community structures (ESCOP, 2001). Agriculture is the economic base of only one-fourth of the rural counties in the United States, and continuing consolidation means that agriculture provides fewer jobs even in those counties; so improvements in agricultural income cannot provide the sole solution to rural economic development. In some regions, dwindling rural economies have caused many once-viable communities to become almost "ghost towns" characterized by declining education and health services and supported by a weakened tax base. In other areas, agriculture remains crucial to rural amenities and quality of life that may ultimately promote a broader base for the rural economy. New social-science research tools enable a better understanding of economic links and of the role of social and human capital, entrepreneurism, and leadership in rural growth. REE research can provide the basis of programs that help people and institutions to respond successfully to continuing economic and institutional change.

## Evaluate the Effects of Changes in Market Structure

A smaller number of worldwide companies now dominate the global agricultural industry, and broad alliances have become common among producers of new technologies, pharmaceuticals, and food (NRC, 2002a). In US agricultural markets, relatively few corporate entities control a major share of sales volume in the input, farming, processing, and food-distribution industries. Moreover, links between sectors are far more common than before. A growing number of commodities are now produced under contracts that specify management techniques or product characteristics.

Those structural changes in agricultural markets have important implications for the welfare of producers and consumers through their influence on rural development, price discovery,[6] access to markets, and industry response to changing consumer demands. Research is needed to understand the effects of vertical integration,[7] contracting,[8] and consolidation[9] on performance of the food system. Expanded knowledge also will be essential for designing new institutions for price discovery and market coordination. Research is needed to understand how structural changes influence agriculture's role in the rural economy and the breadth of economic participation in agriculture. Such research will support REE

---

[6]Price discovery is the process by which buyers and sellers find the price that equates available supply with demand.

[7]Vertical integration (or vertical coordination) is the coordination of stages in the agricultural production and marketing chain under common ownership.

[8]Contracts are agreements between producers and companies or other farmers that specify conditions of production or marketing.

[9]Consolidation is the merging or joining of businesses that produce the same product at the same stage of the production and marketing chain (Tweeten and Flora, 2001).

programs to aid decision-makers in the agricultural sector and USDA action-agency programs for rural economic development.

**Meet the Challenge of Rural Development's Changing Context**

The differences in economic activities in rural areas, the differences in economic circumstances between farm households and the general US population (particularly the rural population), and the differences within the farm population (presence of limited-resource farmers) strengthen a rationale for moving research and outreach focus beyond the traditional large-farm constituency (Lobao, 1990; Lobao and Meyer, 2001; Mishra et al., 2002; USDA, 2001). Solving the problems of rural communities will require research to understand how to broaden and diversify the rural economic base, how to increase access to emerging markets, and how to invest in developing skills for responding to change. Sociologic and economic research must provide the knowledge essential for strengthening community leadership skills, diversifying participation in the global economy, developing new markets, understanding needed economic transitions, meeting the costs of adjusting to change, and eliminating the "digital divide," or gap between those with access to information infrastructure and those without (ESCOP, 2001). Understanding of the roles of social and human capital, entrepreneurism, and leadership in building successful rural communities constitutes a basic social-science research frontier.

Rural communities, farm workers, resource-poor, and small-scale farmers—more likely to be black or female—face unique challenges (Lobao and Meyer, 2001; USDA, 1999a, 2001). Since 1986, farm-labor contractors have taken a much greater role in the functioning of farm-labor markets. This occurred to shift liability away from growers when undocumented farm workers were employed in the fields (Martin et al., 1994). Access to health services, exposure to health risks, such as agricultural chemicals, and children's education are issues of particular concern to them. Research is needed to understand how occupational and geographical mobility, new technologies, new markets, and social programs can benefit these groups. Ultimately, findings from such research would be applied in efforts to strengthen rural communities through participatory decision-making and entrepreneurial economic development.

## ADVANCING THE FRONTIERS

The research frontiers noted above all have to do with agriculture as a system that links many biologic, physical, social, and economic processes. Tomorrow's agricultural research must explicitly identify and address those links so that progress in one agricultural sector does not inadvertently create or exacerbate problems in another. There is a need for systematic research on indicators of the environmental, social, and community impacts of REE agencies (discussed further in Chapter 6). These research opportunities will require greater emphasis on engaging all relevant disciplines in developing workable, effective, and long-term solutions and in providing early assessments of new technologies and policy shifts. An approach that addresses researchable questions on all relevant scales—from genes, fields, and farms to landscapes, watersheds, and regions—also will be essential, as will a long-term strategic approach that looks not only at near-term issues but also at questions best studied over periods of decades or longer (Box 3-1).

Research on the frontiers identified above is often best undertaken in the public sector because many of the challenges will not be fully addressed through private-sector research and because related USDA programs and policies will require an expanded research base (see Box 3-2). In many cases, new research opportunities will require expanded collaboration among scientific disciplines, federal agencies, or international organizations. REE is uniquely positioned to address the new frontiers in agricultural research—alone or in collaboration with other partners—through its historical strengths in mission-oriented research, collaboration at all levels, and responsibility for collecting food and agricultural data.

**RECOMMENDATION 1: REE should provide leadership for the agricultural community in exploring research frontiers in food, health, environment, and communities. REE should build on its historical strengths and become a scientific leader in using new technologies and emerging scientific paradigms to pursue strategic, long-term research goals. A greater emphasis on multidisciplinary work that engages all relevant disciplines will be needed to address many new research frontiers.**

Successfully addressing these research frontiers will require that USDA become a scientific leader in identifying, evaluating, and deploying new technologies and emerging scientific paradigms. The limited disciplinary coverage of each USDA agency and the mixed history of communication among research disciplines pose a serious challenge to advancing this new level of multidisciplinary research. One essential step will be improved coordination among USDA agencies and between USDA and other federal and nonfederal agencies and institutions. An equally critical change will be to move from a narrowly focused set of research priorities to a more strategic and long-term approach to food and agricultural research. Ultimately, that will require either an expansion or a reallocation of research funds across REE because some of the most compelling research needs have received little funding in past agency budgets. Chapter 4 discusses such institutional and resource issues in detail.

## BOX 3-1
## Research on Relevant Spatial and Temporal Scales

**Hypoxia in the Gulf of Mexico**

The decade-long debate about the cause of hypoxia in the Gulf of Mexico is a model illustrating the need for an integrated, multiscale approach to complex agroenvironmental problems. Over the last 20 years, an area of the northern gulf has become anoxic during the summer, driving fish away and suffocating benthic organisms. In 1999, the zone was the size of New Jersey. Since the anoxia was first discovered to be widespread and growing in the late 1980s (Turner and Rabalais, 1991, 1994), marine ecologists have suspected a linkage to inorganic nitrogen entering the gulf from the Mississippi River (Rabalais et al., 1996); nitrogen is known to limit primary productivity in marine systems: higher nitrate concentrations cause algal blooms at the ocean surface. As algae die and sink, they are decomposed by heterotrophic bacteria that also consume oxygen; this results in a water column that is progressively more oxygen-depleted over the course of the growing season. There has been no documented evidence of harm to gulf fisheries—trawlers have moved their operations out to deeper waters—but many are concerned that the so-called dead zone will continue to grow and eventually become a major economic problem (Ferber, 2001). Similar dead zones have been documented off the coasts of Europe, Japan, and other parts of the United States (Diaz and Rosenberg, 1995).

The cause of gulf hypoxia has been hotly debated in the popular and scientific press. Although it is widely acknowledged that the proximate cause is nitrate, where in the watershed the nitrate originates has been actively disputed. Most farm-advocacy and commodity groups have contended that nitrogen discharged from the Mississippi drainage is the result of coastal upwelling or naturally high-nitrogen discharge rates from Mississippi catchments or urban areas in the catchment, in particular from wastewater-treatment plants and overfertilized lawns and golf courses. Biogeochemists point out that mass-balance models for different watersheds in the region and for the watershed as a whole support a much more widely dispersed source—nitrogen leached from fertilized farm fields. In 2000, the National Research Council Committee on Causes and Management for Coastal Eutrophication concluded that excessive use of fertilizer and other farming practices have caused the hypoxic condition (NRC, 2000a), a finding in agreement with the White House Committee on the Environment and Natural Resources *Gulf of Mexico Hypoxia Assessment* (CENR, 2000).

The debate is relevant to a new paradigm for environmental research in agriculture. Identification of the cause of the problem and its eventual solution require a systems approach. Nitrogen fertilizer added to cultivated fields does not simply filter through to surface waters; rather, it undergoes a wide variety of transformations, many of them biologic, before emerging as nitrate at the mouth of the Mississippi. Nitrogen added to cultivated fields is first transformed by bacteria to a form available to plants, and it then is either taken up by plants or left in the soil solution available for leaching or transformation to nitrogen gas by other soil bacteria. If taken up by plants, the nitrogen will be either harvested or left as plant residue on the field, to be recycled again into mineral nitrogen over the following months. Nitrate that is leached out of the field enters groundwater, from which it may emerge years later in streams that may also process it to other forms, including harmless dinitrogen gas.

A wide variety of management decisions and environmental factors affect the rate at which nitrogen undergoes its various transformations. Disciplinary knowledge of rhizosphere uptake rates and hydrologic flow paths are important for understanding potential fates, but a sufficient understanding of nitrate leaching requires integrating the full suite of processes that interact to deposit nitrate in surface waters. Often, quantitative system models can help to identify the most important control points in the system, which can then be verified experimentally. By definition, this understanding requires knowledge in a wide variety of disciplines—soil and aquatic microbiology and chemistry, soil physics, agronomy, plant physiology, hydrology, geochemistry, and others. Placing this disciplinary information into an integrated ecologic context is the hallmark of a sound systems approach.

The hypoxia challenge also illustrates the need for long-term research on multiple spatial scales. Small pockets of hypoxia were first noted in the gulf in the early 1970s; its growth has been a long-term phenomenon that will probably take as long to fully understand and halt—especially given recent evidence that nitrate leached from Illinois fields may take 2 decades to emerge in surface water. And although the leached nitrate is a problem mainly on the large, aggregated scale of the entire Mississippi watershed, its disaggregated causes—and possibly its solutions—can be traced back to the microscopic interactions among bacteria, root hairs, and soil particles that regulate the uptake of individual fertilizer molecules.

Finally, the eventual solution to the hypoxia problem must necessarily involve social scientists and extension efforts. The policy options needed to induce producers to conserve nitrogen will be as important to understand as the biophysical causes of nitrogen loss from their fields. And enacting these policies and promoting new best-management practices will require effective education and demonstration efforts on the part of agricultural extension.

*continued*

## BOX 3-1 Continued

Ecology in general and biogeochemistry in particular were not sufficiently advanced 30 years ago to have anticipated the current magnitude of this problem. One might reasonably expect, however, that the emerging environmental-research paradigm for agriculture, with its focus on an integrated understanding of systems on multiple scales, will catch these sorts of problems at an earlier stage and provide options for their solution before major economic harm.

**Long-Term Basic Research in Environmental Science**

The National Science Foundation (NSF) Long-Term Ecological Research (LTER) Program is a good example of the value of long-term basic research for addressing unforeseen environmental problems. The success of the program (Kaiser, 2001) stems largely from its focus on building a basic, long-term, multiple-scale understanding of ecologic interactions in specific ecosystems. NSF provides about $700,000 per year to each of 24 sites to support research in five core subjects—primary production, populations, nutrient cycling, organic-matter dynamics, and disturbance. Research foci differ by site, and each site actively hosts external projects funded by traditional short-term funding mechanisms, including the US Department of Agriculture National Research Initiative. At a few LTER sites, REE agencies are also involved via modest funding from state agricultural experiment stations (Cooperative State Research, Education, and Extension Service) and shared site support (Agricultural Research Service).

Although sites are focused largely on examinations of basic ecologic processes in unmanaged ecosystems—only one site includes field crops, and only three include livestock—there are many examples of research results with relevance to agriculture:

- At the Konza Prairie site in Kansas, researchers have discovered the importance of the interaction between disturbances generated by fire and grazing for maintaining grassland biodiversity (Collins et al., 1998). Ten-year experiments showed that fire or grazing alone is insufficient for maintaining plant species diversity in tallgrass prairie—that grazing in addition to fire is needed to maintain species richness.
- At the Short Grass Steppe site in Colorado, analysis of a 23-year dataset of plant productivity showed that long-term increases in average minimum temperatures decreased primary production by the dominant $C_4$ grass and increased production by native and exotic forbs (Alward et al., 1999). Thus, long-term climate change

appears to be making grasslands more vulnerable to invasion by exotic species and less tolerant of drought and grazing.
- An experiment involving nutrient additions to 12 headwater streams at different LTER sites (Peterson et al., 2001) demonstrated the remarkable capacity of small streams to remove nitrogen from incoming catchment water. Results suggest that small streams, highly vulnerable to channelization and elimination in agricultural landscapes, are crucial for regulating water chemistry in large drainages because their large surface-to-volume ratios favor rapid nitrogen uptake and processing.
- At the Cedar Creek LTER site in Minnesota, researchers have documented an association between plant diversity and primary productivity (Tilman et al., 2001). In a 7-year experiment, multi-species grassland plots outperformed the best grassland monocultures, establishing a clear linkage between biodiversity and ecosystem productivity.
- At the Kellogg Biological Station LTER site in Michigan, researchers have used 10 years of soil-carbon and greenhouse-gas measurements to contrast the global-warming potentials of different cropped and unmanaged ecosystems (Robertson et al., 2000). The analysis found multiple opportunities for modern agriculture to contribute to greenhouse-gas mitigation in addition to the well-established capacity for no-till and cover-cropped soils to sequester carbon.

Long-term environmental field research can yield critical insights not only for the field of ecology but also in such fields as soil science and agronomy (e.g., Paul et al., 1997; Rasmussen et al., 1998). Indeed, the longest continuously measured field experiments in existence are those at the Rothamsted Agricultural Experiment Station in England (Jenkinson, 1991). Today, long-term agronomic research sites in the United States are rare and poorly funded. Yet long-term field studies will be particularly important for research that anticipates the environmental impacts of shifting agricultural practices and identifies mechanisms for agriculture to deliver diverse environmental benefits.

## BOX 3-2
## Finding Resources to Explore Research Frontiers

Historically, agriculture has been unique in its need for publicly funded research because of the structure of firms, the unique natural risks asso-

*continued*

## BOX 3-2 Continued

ciated with production that influence the adoption process, the nonexcludable nature of many agricultural technologies, and the spillover benefits from technology beyond the primary users. Public agricultural research is needed because the public sector can undertake research that is long-term, large-scale, and risky and provides benefits that cannot be appropriated by individual firms (Alston and Pardey, 1996). Because much agricultural research is location-specific, there is a broadly defined division of labor between federal and state research, so state experiment stations focus on state or regional issues, and federal research addresses national issues (Huffman and Evenson, 1993). Because the division between state and national or between public and private is never entirely discrete, some overlap will occur, and collaborations among state, federal, and private entities have complementarities (see Chapter 5).

The need for public agricultural research continues, but the division of labor is shifting. The structure of agriculture is changing, the spillovers from research are now international in scope, and the incentives for private research investment are increasing (Reilly and Schimmelpfennig, 2000). Federal public research should therefore become even more focused on

- Research that provides new "platforms" of discovery for multiple private and/or local applications.
- Research with broad public benefits, where the returns to research cannot be appropriated by private developers of technology. Such benefits as enhanced public health or environmental services are often more difficult for private firms to capture, and thus these are important goals for public research.
- Research that addresses national needs and provides benefits that are shared widely.

The research frontiers identified in this chapter would all provide substantial public benefits, but within these frontiers the REE agencies will need to establish how they can best contribute either through basic research or through partnerships with universities and the private sector.

It must also be recognized that research may or may not provide solutions to pressing social, economic, health, and environmental concerns. Research takes place in the context of other public policies and private actions that influence how research results are used. Policies to address society's concerns will create incentives for innovation, which may in turn influence the appropriate public research role. Thus, as we discuss further in Chapters 4, 5, and 6, the process of defining the public role and strategy within these research frontiers must be continuing and dynamic.

## SUMMARY

This chapter has identified new research directions related to five major phenomena: globalization; the emergence of pathogens; links between diet, health promotion, and disease prevention; the relationship between agriculture and the environment; and changes in rural communities. A multidisciplinary, strategic approach, consideration of relevant spatial and temporal scales, and coordination with other agencies will be essential for addressing many of those research topics. A unique role for the public sector, and specifically for REE, in undertaking the research is justified, given the expanded research needs of USDA programs and policies and the limited capability for private-sector research to address it.

## REFERENCES

Alston, J.A., and P.G. Pardey. 1996. Making Science Pay: The Economics of Agricultural R&D Policy. Washington, DC: American Enterprise Institute for Public Policy Research.

Alward, R.D., J.K. Detling, and D.G. Milchunas. 1999. Grassland vegetation changes and nocturnal global warming. Science 283:229–231.

Babbitt, B. 1998. Statement by Secretary of the Interior on invasive alien species. Pp. 8–10 in National Weed Symposium. Denver, CO: Bureau of Land Management.

Brown, P., S. Gettner, and C. Somerville. 1999. Genetic engineering of plant lipids. Annual Review of Nutrition 19:197–216.

Buckley, D.H., and T.M. Schmidt. 2001a. The structure of microbial communities in soil and the lasting impacts of cultivation. Microbial Ecology 42:11–21.

Buckley, D.H., and T.M. Schmidt. 2001b. Exploring the biodiversity of soil—A microbial rainforest. Pp. 183–208 in Biodiversity of Microbial Life: Foundation of Earth's Biosphere, J.T. Staley and A.L Reeysenbach, eds. New York: John Wiley and Sons.

Burn, P., and G. Kishore. 2000. Food as a source of health enhancing compounds. AgBioForum 3(1):3–9. Available online at *http://www.agbioforum.org*.

Byers, J. E., S. Reichard, J.M. Randall, I.M. Parker, C.S. Smith, W.M. Lonsdale, I.A.E. Atkinson, T.R. Seastedt, M. Williamson, E. Chornesky, and D. Hayes. 2002. Directing research to reduce the impacts of nonindigenous species. Conservation Biology 16(3):630–640.

CAST (Council on Agricultural Science and Technology). 2000. Storing Carbon in Agricultural Soils to Help Mitigate Global Warming. N.J. Rosenberg and R.C. Izaurralde, eds. IP14, April. Washington, DC: Battelle Pacific Northwest Laboratory.

CAST (Council on Agricultural Science and Technology). 2003. Agriculture's Response to Global Climate Change. B. Babcock and K. Paustian, eds. CAST Report No. 138. Ames, IA: Council on Agricultural Science and Technology.

CENR (Committee on Environment and Natural Resources, National Science and Technology Council). 2000. Integrated Assessment of Hypoxia in the Northern Gulf of Mexico. Washington, DC: National Science and Technology Council Committee on Environment and Natural Resources.

Childs, N.M. 2001. Marketing issues for functional foods and nutraceuticals. In Handbook of Nutraceuticals and Functional Foods, R.C. Wildman, ed. New York: CRC Press.

Collins, S.L., A.K. Knapp, J.M. Briggs, J.M. Blair, and E.M. Steinauer. 1998. Modulation of diversity by grazing and mowing in native tallgrass prairie. Science 280:745–747.

Delgado, C., M. Rosegrant, H. Steinfeld, S. Ehui, and C. Courbois. 2001. Livestock to 2020: The next food revolution. Pp. 89–94 in The Unfinished Agenda, P. Pinstrup-Andersen and R. Pandya-Lorch, eds. Washington, DC: International Food Policy Research Institute.

Della Penna, D. 1999. Nutritional genomics: Manipulating plant micronutrients to improve human health. Science 285:375–379.
Diaz, R.J., and R. Rosenberg. 1995. Marine benthic hypoxia: A review of its ecological effects and the behavioral responses of benthic macrofauna. Pp. 245–303 in Oceanography and Marine Biology, A.D. Ansell, R.N. Gibson, and M. Barnes, eds. Oban, Argyll, Scotland: UCL Press.
DiTomaso, J.M. 2000. Invasive weeds in rangelands: Species, impacts, and management. Weed Science 48:255–265.
Duncan, R.C. 2001. World energy production, population growth, and the road to the Olduvai Gorge. Population and Environment 22:503–522.
Eisenberg, D.M., R.B. Davis, S.L. Ettner, S. Appel, S. Wilkey, M. Van Rompay, and R.C. Kessler. 1998. Trends in alternative medicine use in the United States, 1990–1997. Journal of the American Medical Association 280:1569–1575.
ESCOP (Experiment Station Committee on Organization and Policy). 2001. A Science Roadmap for Agriculture. National Association of State Universities and Land-Grant Colleges. Available online at *www.nasulgc.org/publications/agriculture/science%20Roadmap2.pdf*.
Ewel, J. J., D.J. O'Dowd, J. Bergelson, C. Daehler, C.M. D'Antonio, L.D. Gomez, D.R. Gordon, R.J. Hobbs, A. Holt, K.R. Hooper, C.E. Hughes, M. LaHart, R.R.B. Leakey, W.G. Lee, L.L. Loope, D.H. Lorence, S.M. Louda, A.E. Lugo, and P.B. McEvoy. 1999. Deliberate introductions of species: Research needs. Bioscience 49(8):619–630.
FASEB LSRO (Federation of American Societies for Experimental Biology, Life Sciences Research Office). 1990. Core indicators of nutritional state for difficult to sample populations. Journal of Nutrition 120(Supplement):1559–1600.
FDA (Food and Drug Administration). 1992. Statement of Policy: Foods Derived from New Plant Varieties. 57 Federal Register 22984.
Feller, I., C.P. Ailes, an J.D. Roessner. 2002. Impacts of research universities on technological innovation in industry: Evidence from engineering research centers. Research Policy 31:457–474.
Ferber, D. 2001. Keeping the Stygian waters at bay. Science 291:969–969.
Frist, W. 2002. When Every Moment Counts: What You Need to Know about Bioterrorism from the Senate's Only Doctor. Lanham, MD: Rowman & Littlefield.
Goolsby, D.A., W.A. Battaglin, B.T. Aulenbach, and R.P. Hooper. 2000. Nitrogen flux and sources in the Mississippi River basin. Science and the Total Environment 248:75–86.
Heinz Center. 1999. Designing a Report on the State of the Nation's Ecosystems: Selected Measures of the Condition of Croplands, Forests, and Coasts and Oceans. Washington, DC: The H. John Heinz III Center for Science, Economics, and the Environment. Available online at *http://www.us-ecosystems.org/croplands/index.html*.
Heller, M.A., and R.S. Eisenberg. 1998. Can patents deter innovation? The anticommons in biomedical research. Science 280:698–701.
Huffman, W., and R. Evenson. 1993. Science for Agriculture: A Long-Term Perspective. Ames, IA: Iowa State University Press.
Hughes, J.M. 2001. Emerging infectious diseases: A CDC perspective. Emerging Infectious Disease 7(3 Supplement):494–496.
IFT (Institute of Food Technologists). 2002. IFT Expert Report on Emerging Microbiological Food Safety Issues: Implications for Control in the 21st Century. Chicago: Institute of Food Technologists.
IPCC (Intergovernmental Panel on Climate Change). 2000. Land use, land-use change, and forestry, R.T. Watson, I.R. Noble, B. Bolin, N.H. Ravindranath, D.J. Verardo, and D.J. Dokken, eds. Cambridge, UK: Cambridge University Press.

IPCC (Intergovernmental Panel on Climate Change). 2001. Climate Change 2001: The Scientific Basis. Contribution of Working Group I to the Third Assessment Report of the Intergovernmental Panel on Climate Change (IPCC), J.T. Houghton, Y. Ding, D.J. Griggs, M. Noguer, P.J. van der Linden, and D. Xiaosu, eds. Cambridge, UK: Cambridge University Press.

Jenkinson, D.S. 1991. The Rothamsted long-term experiments: Are they still of use? Agronomy Journal 83:2–10.

Kaiser, J. 2001. An experiment for all seasons. Science 293:624-627.

Kipnis, V., A.F. Subar, D. Midthune, L.S. Freedman, R. Ballard-Barbash, R. Troiano, S. Bingham, D.A. Schoeller, A. Schatzkin, and R.J. Carroll. (in press). The structure of dietary mesurement error: Results of the OPEN biomarker study. American Journal of Epidemiology.

Kleese, R.A. 2000. Designing crops for added value: A vision, a mission. Pp. 1–10 in Designing Crops for Added Value, C.F. Murphy and D.M. Peterson, eds. Agronomy Series No. 40. Madison, WI: American Society of Agronomy, Crop Science Society of America, Soil Science Society of America.

Knudson, W.A. 2001. Public research universities and the private sector: Engine of economic growth or captive of special interests? Choices, First Quarter 2001: 31–35.

Krauss, R.M., P.T. Williams, F.T. Lindgren, and P.D. Wood. 1988. Coordinate change in levels of human serum low and high-density lipoprotein subclasses in healthy men. Arteriosclerosis 8:155–162.

Kuzminski, L.N. 1994. To the "Forum on Meeting the Challenge: Health, Safety, and Food for America." Briefing Paper Sponsored by the Office of Science and Technology Policy, Executive Office of the President; and the National Science and Technology Council's Committee on Safety, Health, and Food Research and Development. Institute of Food Technologists/US Food Experts Alliance for Strategic Technology (IFT/USFEAST) Steering Committee, November 21.

Lal, R. 1998. Soil erosion impact on agronomic productivity and environment quality. Critical Reviews in Plant Sciences 17(4):319–464.

Lobao, L.M. 1990. Locality and inequality: Farm and industry structure and socioeconomic conditions. Albany, NY: State University of New York Press.

Lobao, L., and K. Meyer. 2001. The great agricultural transition: Crisis, change, and social consequences of twentieth century US farming. Annual Review of Sociology 27:103–124.

Mack, R.N., D. Simberloff, W.M. Lonsdale, H. Evans, M. Clout, and F.A. Bazzaz. 2000. Biotic invasions: Causes, epidemiology, global consequences, and control. Ecological Applications 10:689–710.

Martin, P.L., W.E. Huffman, R.E. Emerson, G.E. Taylor, and R. Rochin. 1994. The Immigration and Control Act of 1986: Implications for Immigrant Farm Labor. Oakland, CA: University of California, Department of Natural Resources.

McNeely, J.A., H.A. Mooney, L.E. Neville, P. Schei, and J.K. Waage, eds. 2001. Global Strategy on Invasive Alien Species. Gland, Switzerland, and Cambridge, UK: IUCN (World Conservation Union) on behalf of the Global Invasive Species Programme.

Mishra, A.K., H.S. El-Osta, M.J. Morehart, J.D. Johnson, and J.W. Hopkins. 2002. Income, wealth, and the economic well-being of farm households. ERS Agricultural Economic Report No. AER 812. 77 pp. July, 2002. Washington, DC: Economic Research Service, US Department of Agriculture. Available online at *http://www.ers.usda.gov/publications/aer812/*.

Mokdad, A.H., B.A. Bowman, E.S. Ford, F. Vinicor, J.S. Marks, and J.P. Koplan. 2001. The continuing epidemics of obesity and diabetes in the United States. Journal of the American Medical Association 286(10):1195–1200.

NISC (National Invasive Species Council). 2002. National Invasive Species Management Plan. Available online at *http://invasivespecies.gov/council/nmp.shtml*.

Nordlee, J.A., S.L. Taylor, J.A. Townsend, L.A. Thomas, and R.K. Bush. 1996. Identification of a brazil-nut allergen in transgenic soybeans. New England Journal of Medicine 334:688–692.

NRC (National Research Council). 1999. Review of the Research Strategy for Biomass-Derived Transportation Fuels. Washington, DC: National Academy Press.
NRC (National Research Council). 2000a. Clean Coastal Waters: Understanding and Reducing the Effects of Nutrient Pollution. Washington, DC: National Academy Press.
NRC (National Research Council). 2000b. Genetically Modified Pest-Protected Plants. Washington, DC: National Academy Press.
NRC (National Research Council). 2000c. The Future Role of Pesticides in US Agriculture. Washington, DC: National Academy Press.
NRC (National Research Council). 2002a. Publicly Funded Agricultural Research and the Changing Structure of US Agriculture. Washington, DC: National Academy Press.
NRC (National Research Council). 2002b. The Scientific Basis for Estimating Air Emissions from Animal Feeding Operations. Interim Report. Washington, DC: National Academy Press.
Parker, D., F. Castillo, and D. Zilberman. 2001. Public-private sector linkages in research and development: The case of US agriculture. American Journal of Agricultural Economics 83(3):736–741.
Pattersson, O. 1997. Pesticide use in Swedish agriculture: The case of a 75 percent reduction. Pp. 79–102 in Techniques for Reducing Pesticides Use: Economic and Environmental Benefits, D. Pimentel, ed. Chichester, UK: John Wiley & Sons.
Paul, E.A., K.A. Paustian, E.T. Elliot, and C.V. Cole, eds. 1997. Soil Organic Matter in Temperate Ecosystems: Long-Term Experiments in North America. Boca Raton, FL: Lewis/CRC Publishers.
Peterson, B.J., W.F. Wollheim, P.J. Mulholland, J.R. Webster, J.L. Meyer, J.L. Tank, E. Marti, W.B. Bowden, H.M. Valett, A.E. Hershey, W.H. McDowell, W.K. Dodds, S.K. Hamilton, S. Gregory, and D.D. Morrall. 2001. Control of nitrogen export from watersheds by headwater streams. Science 292:86–90.
Pimentel, D. 2001. The limitations of biomass energy. In Encyclopedia of Physical Sciences and Technology. San Diego, CA: Academic Press.
Pimentel, D., L. McLaughlin, A. Zepp, B. Kakitan, T. Kraus, P. Kleinman, F. Vancini, W.J. Roach, E. Graap, W.S. Keeton, and G. Selig. 1993. Environmental and economic effects of reducing pesticide use in agriculture. Agriculture, Ecosystems, and Environment 46(1–4):273–288.
Pimentel, D., C. Harvey, P. Resosudarmo, K. Sinclair, D. Kurz, M. McNair, S. Crist, L. Sphritz, L. Fitton, R. Saffouri, and R. Blair. 1995. Environmental and economic costs of soil erosion and conservation benefits. Science 267:1117–1123.
Pimentel, D., R. Doughty, C. Carothers, S. Lamberson, N. Bora, and K. Lee. 2002. Energy inputs in crop production: Comparison of developed and developing countries. Pp. 129–151 in Food Security and Environmental Quality in the Developing World, L. Lal, D. Hansen, N. Uphoff, and S. Slack, eds. Boca Raton, FL: CRC Press.
Rabalais, N.N., R.E. Turner, N. Justic, Q. Dortch, W.J. Wiseman, and B.K. Sen Gupta. 1996. Nutrient changes in the Mississippi River and system responses on the adjacent continental shelf. Estuaries 19:386–407.
Rasmussen, P.E., K.W.T. Goulding, J.R. Brown, P.R. Grace, H.H. Janzen, and M. Korschens. 1998. Long-term agroecosystem experiments: Assessing agricultural sustainability and global change. Science 282:893–896.
Reilly, J.M., and D.E. Schimmelpfennig. 2000. Public-private collaboration in agricultural research: The future. In Public-Private Collaboration in Agricultural Research: New Institutional Arrangements and Economic Implications, K.O. Fuglie and D.E. Schimmelpfennig, eds. Ames, IA: Iowa State University Press.
Robertson, G.P., and R.R. Harwood. 2001. Sustainable agriculture. Pp. 99–108 in Encyclopedia of Biodiversity, S.A. Levin, ed. New York: Academic Press.
Robertson, G.P., E.A. Paul, and R.R. Harwood. 2000. Greenhouse gases in intensive agriculture: Contributions of individual gases to the radiative forcing of the atmosphere. Science 289:1922–1925.

Smith, K.R., N. Ballenger, K. Day-Rubenstein, P. Heisey, and C. Klotz-Ingram. 1999. Biotechnology research: Weighing the options for a new public-private balance. Pp. 22–25 in Agricultural Outlook, October. Washington, DC: Economic Research Service, US Department of Agriculture.

Smith, V.H., G.D. Tilman, and J.C. Nekola. 1999. Eutrophication: Impacts of excess nutrient inputs on freshwater, marine, and terrestrial ecosystems. Environment and Pollution 100:179–196.

Subar, A.F., V. Kipnis, R.P. Troiano, D. Midthune, D.A. Schoeller, S. Bingham, C.O. Sharbaugh, J. Trabulsi, S. Runswick, R. Ballard-Barbash, J. Sunshine, and A. Schatzkin. (in press). Using intake biomarkers to evaluate the extent of dietary misreporting in a large sample of adults: The Observing Protein and Energy Nutrition (OPEN) study. American Journal of Epidemiology.

Sullivan, A.F., and E. Choi. 2002. Hunger and Food Insecurity in the Fifty States: 1998–2000. Waltham, MA: Center on Hunger and Poverty, Heller School for Social Policy and Management, Brandeis University. Available online at *http://www.centeronhunger.org/pdf/statedata98-00.pdf.*

Tilman, D., P.B. Reich, J. Knops, D. Wedin, T. Mielke, and C. Lehman. 2001. Diversity and productivity in a long-term grassland experiment. Science 293:843–845.

Todd, E. 2001. Epidemiology and globalization of foodborne disease. Pp. 1–22 in Guide to Foodborne Pathogens, R.G. Labbé and S. Garcia, eds. New York: John Wiley and Sons.

Turner, R.E., and N.N. Rabalais. 1991. Changes in Mississippi River water quality this century. BioScience 41:140–147.

Turner, R.E., and N.N. Rabalais. 1994. Coastal eutrophication near the Mississippi River delta. Nature 368: 619–621.

Tweeten, L.G., and C.B. Flora. 2001. Vertical Coordination of Agriculture in Farming-Dependent Areas. Task Force Report No. 137. Ames, IA: Council for Agricultural Science and Technology.

Unnevehr, L.J., and T. Roberts. 2002. Food safety incentives in a changing world food system. Journal of Food Control 13:73–76.

USBC (United States Bureau of the Census). 1999. Statistical Abstract of the United States 1999. Washington, DC: US Government Printing Office.

US Congress. 2002. P.L. (Public Law) 107-171. Farm Security and Rural Investment Act of 2002. H.R. 2646.

USDA (US Department of Agriculture). 1999a. Census of Agriculture. Washington, DC: National Agricultural Statistics Service, US Department of Agriculture.

USDA (US Department of Agriculture). 1999b. US Action Plan on Food Security. Washington, DC: Foreign Agricultural Service, US Department of Agriculture.

USDA (US Department of Agriculture). 2000a. Changes in Average Annual Soil Erosion by Water on Cropland and CRP Land, 1992–1997. Natural Resources Conservation Service. US Department of Agriculture. Revised December 2000.

USDA (US Department of Agriculture). 2000b. Changes in Average Annual Soil Erosion by Wind on Cropland and CRP Land, 1992–1997. Natural Resources Conservation Service. US Department of Agriculture. Revised December 2000.

USDA (US Department of Agriculture). 2001. Structural and Financial Characteristics of U.S. Farms: 2001 Family Farm Report. R.A. Hoppe, ed. Agriculture Information Bulletin No. 768. Washington, DC: Resource Economics Division, Economic Research Service, US Department of Agriculture.

USDHHS (US Department of Health and Human Services). 1970. Centers for Disease Control and Prevention. National Center for Health Statistics, National Health Examination Survey, 1963–1970. Overweight Among US Children and Adolescents. Available online at *http://www.cdc.gov/nchs/about/major/nhanes/databriefs/overwght.pdf.*

USDHHS (US Department of Health and Human Services). 1974. Centers for Disease Control and Prevention. National Center for Health Statistics. National Health and Nutrition Examination Survey I, 1971–1974. Overweight Among US Children and Adolescents. Available online at *http://www.cdc.gov/nchs/about/major/nhanes/databriefs/overwght.pdf.*

USDHHS (US Department of Health and Human Services). 1980. Centers for Disease Control and Prevention. National Center for Health Statistics. National Health and Nutrition Examination Survey II. Prevalence of Overweight and Obesity Among Adults: United States, 1976–1980. Available online at *http://www.cdc.gov/nchs/products/pubs/pubd/hestats/obese/obse99t2.htm.*

USDHHS (US Department of Health and Human Services). 1988. Centers for Disease Control and Prevention. National Center for Health Statistics. National Health and Nutrition Examination Survey III. Prevalence of Overweight and Obesity Among Adults: United States, 1988–1994. Available online at *http://www.cdc.gov/nchs/products/pubs/pubd/hestats/obese/obse99t2.htm.*

USDHHS (US Department of Health and Human Services). 1999. Centers for Disease Control and Prevention. National Center for Health Statistics. National Health and Nutrition Examination Survey. Prevalence of Overweight and Obesity Among Adults: United States, 1999. Available online at *http://www.cdc.gov/nchs/products/pubs/pubd/hestats/obese/obse99.htm.*

USDHHS (US Department of Health and Human Services). 2001. Centers for Disease Control and Prevention. Investigation of Human Health Effects Associated with Potential Exposure to Genetically Modified Corn. A Report to the US Food and Drug Administration from the Centers for Disease Control and Prevention. June 11. Washington, DC: US Department of Health and Human Services, Centers for Disease Control and Prevention.

USEPA (US Environmental Protection Agency). 1998. Pesticide Fact Sheet: *Bacillus thuringiensis* subspecies tolworthi Cry 9C Protein and the Genetic Material Necessary for its Production in Corn. Issued May.

USGS (US Geological Survey). 1999. The Quality of Our Nation's Waters: Nutrients and Pesticides. US Geological Survey Circular 1225. Available online at *http://water.usgs.gov/pubs/circ/circ1225/.*

Van Elswyk, M.E., S.D. Hatch, G.G. Stella, P.K. Mayo, and K.S. Kubena. 1998. Poultry-based alternatives for enhancing the omega-3 fatty acid content of American diets, A.P. Simopoulos, ed. The return of omega-3 fatty acids into the food supply. World Review of Nutrition and Dietetics 83:102–115.

Wilcove, D.S., D. Rothstein, J. Dubow, A. Phillips, and E. Losos. 1998. Quantifying threats to imperiled species in the United States. BioScience 48:607–615.

# 4

# Setting the Research Strategy

To make progress toward the research frontiers identified in Chapter 3, the US Department of Agriculture (USDA) Research, Education, and Economics (REE) mission area will need to be responsive to and direct research toward the changing context and role of US agriculture. Examples of that change are agriculture's broadening scope, shifting opportunities in world markets, new scientific discoveries and paradigms, and the private sector's expanding research efforts. This chapter addresses REE's ability to respond to change by exploring its capacity for setting priorities, for making discretionary changes in resource allocation, and for understanding and working with a broad array of stakeholders.

Many cross-cutting, complementary, and contradictory forces help shape priorities and resource allocations for agricultural research and education. There are problems to solve, stakeholders to serve, agencies to collaborate with, and knowledge to generate and disseminate. Authorization and appropriation legislation lays the foundation for focus and funding. Congressional earmarks annually add a set of research projects identified by largely political rather than scientific criteria. Each incoming administration brings its own initiatives to the table. Finally, USDA and the REE agencies develop broad goals and objectives through periodic strategic planning and performance reporting mandated by the Government Performance and Results Act (GPRA) (US Congress, 1993). Stakeholder input drives REE priority-setting at all levels through congressional lobbying and through mechanisms established in the 1998 Agricultural Research, Extension, and Education Reform Act (AREERA) (US Congress, 1998). The annual budget process wraps together processes and priorities that emerge at all four levels—Congress, the administration, the department, and agencies.

## FUNDING SOURCES AND TRENDS

Appropriations by the US Congress set the bounds for research conducted by REE agencies. Although those appropriations have grown in synchrony with other nondefense research expenditures, their share of overall support for agricultural research in the United States has declined as private and state research investments have increased. Congress and the executive branch routinely circumscribe the direction and use of federal funds for agricultural research through earmarking and other means.

### Trends in Federal Funding of Agricultural Research

Total REE funding was almost constant in 2000 dollars from 1985 to 2001 (Figure 4-1; Appendix Table F-1). Funding for the Agricultural Research Service (ARS) grew slightly by 2001; however, real funding for the other agencies, particularly the Economic Research Service (ERS), declined slightly. The Cooperative State Research, Education, and Extension Service (CSREES) funding varied, with a large decline in 1998 resulting from discontinuation of facility funding and a large increase in 2000 resulting from funding increases to the Initiative for Future Agriculture and Food Systems (IFAFS) and Fund for Rural America (FRA) programs.

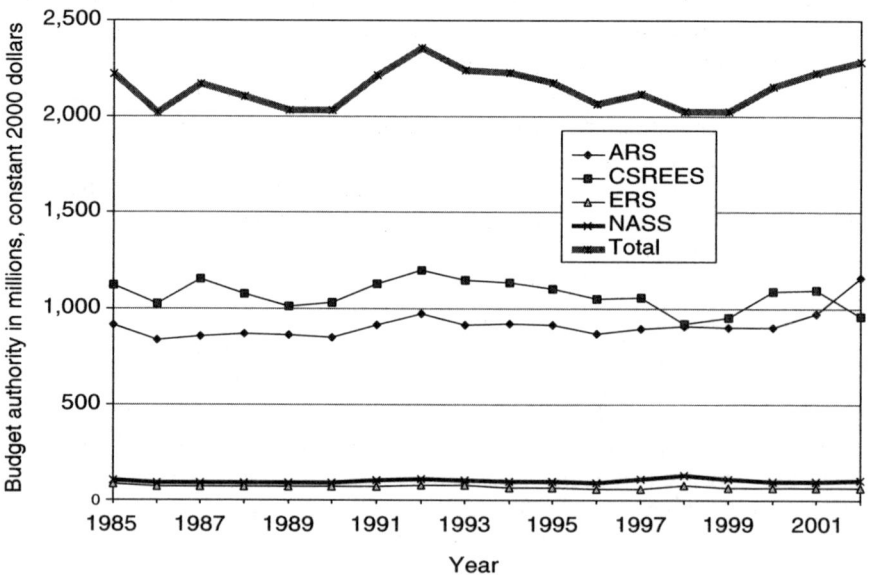

**FIGURE 4-1** Research, Education, and Economics budget authority by agency for FY 1985–2001 and 2002 estimate, in constant 2000 dollars. Source: USDA Office of Budget and Program Analysis (USDA, 2002a).

USDA is part of a much larger and complex R&D system. Assessing the significance of the REE funding levels requires consideration of other federal research agencies and total public-sector (including state) and private-sector expenditures for agricultural research.

USDA research appropriations have grown at about the same average rate as total nondefense research expenditures since 1976 in the federal government, but growth has slowed since 1996. Appendix Table F-2 shows federal research expenditures by agency over time in constant 2000 dollars (AAAS, 2002). USDA research expenditures increased by 24% from 1976 to 2001. Over the same period, the percentage rate of budget growth was 157% at the National Institutes of Health (NIH) and 46% at the National Science Foundation (NSF). The USDA share of federal nondefense expenditures peaked in 1986 at 5.7% but has since declined, reaching its lowest point in 16 years in 2001 (4.8%). In comparison, the NIH FY 2001 budget accounted for 43% of the nondefense budget, and the NSF FY 2001 budget accounted for 7.3% of the nondefense budget. Above-average growth in nondefense agencies, such as NIH and NSF, accounts for part of the change, reflecting society's growing concern with health and increased willingness to fund basic research. If NIH funding, which skews other comparisons, is excluded from the FY 2001 nondefense R&D budget, USDA accounts for nearly 10% of the total nondefense budget and is the fourth-largest supporter of R&D, after the National Aeronautics and Space Administration (NASA), the Department of Energy (DOE), and NSF. Despite the decline, USDA remains the sixth-largest supporter of R&D in the federal government and supports 7% of federal research in the life sciences and 11% in the social sciences (NSF, 2002b). Total expenditures for all public and private agricultural research were roughly $8.5 billion in 1998 (2000 constant dollars), of which $4.9 billion was in the private sector and $3.6 billion in the public sector. REE resources were about $2 billion of the $3.6 billion. Private agricultural R&D expenditures have grown at more than twice the rate of public agricultural research expenditures over the last decade (Figure 4-2; Appendix Table F-3).

USDA is an integral part of the agricultural-research system at the state level because it provides research support to the state agricultural experiment stations (SAESs). State recipients of the funds include not only the SAESs but also the 1890 universities, the schools of forestry, the colleges of veterinary medicine, and other cooperating institutions (which are few). In 2000, of the total of $365 million of research funding from REE, 80% went to the SAESs, 9% to the 1890 universities,[1] 2.7% to the schools of forestry, 1.9% to the veterinary colleges, and 6% to other institutions (USDA, 2000d). On the average, the SAESs fared better

---

[1] The 1890s institutions were created as a result of the Second Morrill Act of 1890 (US Congress, 1890), expanding the 1862 system of land-grant universities to include African American institutions. There are 17 1890s institutions—including one private institution, Tuskegee University—located primarily in the Southeast.

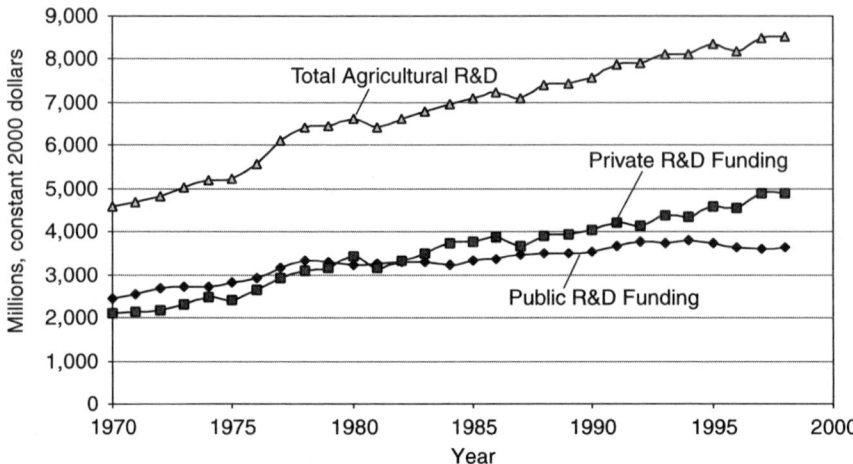

**FIGURE 4-2** Total public and private expenditures, 1970–1998, in constant 2000 dollars. Source: USDA-CRIS Inventory of Agricultural Research (various years); Available at *http://www.ers.usda.gov/data/agresearchfunding/*; updated from Klotz et al. (1995).

over the last 2 decades than the USDA research institutions, growing from $1.89 billion (in constant 2000 dollars) in research funding in 1980 to $2.23 billion in 2000 (Appendix Table F-4). Funding of intramural research institutions declined from $948 million in 1980 to $870 million in 2000 (Appendix Table F-5). The growth in funding of the SAESs resulted in part from increases in non-USDA federal contracts and grants, which now make up 12.8% of total SAES research funding and approach the level of support received through all USDA funding mechanisms. Increased support of the SAESs from industry, commodity groups, and foundations also contributed to the growth in funding. USDA support is thus playing a declining role in SAES funding, having declined to 16.5% of the total from 28.4% in 1980 and 26.1% in 1990. In the future, vision and leadership would be required for USDA to influence the research directions of the SAESs, given that it contributes only a small share of total SAES research funding.

It may be useful to put US agricultural-research investment into international perspective as well. One indicator of the commitment to public agricultural research is the ratio of agricultural R&D to national agricultural output, as measured by agricultural GDP; this indicates the intensity of investment in agricultural research. The intensity of US investment in public agricultural research can be compared with that in other developed countries. In 1993, the United States spent $2.45 on public agricultural R&D for every $100 of agricultural output, ranking last when compared with the United Kingdom (2.9), New Zealand (3.09), Australia (3.66), The Netherlands (3.92), and the average of all developed countries (2.75) in agricultural-research intensity (Pardey et al., 1999).

FINDING: Federal nondefense expenditures have grown faster for health research and for basic research than for agricultural research, and private-sector expenditures for agricultural research now exceed public-sector expenditures. Funds from USDA are declining in importance in SAES funding. The US investment in agricultural research relative to agricultural GDP is below average among all developed countries. The department faces increased challenges in providing leadership for agricultural R&D, in being strategic in use of its limited funds, and in realizing the complementary benefits of agricultural research with other publicly funded research.

### Earmarking, Special Grants, and National Initiatives

Congressional earmarks for research funding play a special role in determining how research resources are allocated (Huffman and Evenson, 1993). Earmarks include projects, facilities, instruments, or other academic or research-related items that are directly funded by Congress. Members of Congress generally identify earmarks in response to lobbying by academic institutions, individual researchers, or other special interests. Budgets appropriated to REE in FY 2002 included $225 million (10% of the total budget) in earmarks. In FY 2002, earmarks appropriated to ARS (about 174) amounted to $89 million, and earmarks appropriated to CSREES (about 246) $136 million (US Congress, 2001a, 2001b, 2001c, 2002a; USDA, 2001b). Congressional earmarks since 1993 for ARS research and CSREES research, education, and extension appear to have increased, as summarized in Appendix Tables F-10a and F-10b. The ERS and National Agricultural Statistics Service (NASS) budgets included no earmarks.[2]

In contrast to federally funded agricultural research, federally funded basic research and health research are not heavily earmarked. NSF carries very few earmarks. In FY 2002, for example, NSF reported two earmarks totaling $50 million, or 1% of the total budget of $4.8 billion. NIH also reported very few earmarks. In FY 2002, for example, only the National Center for Research Resources, the Office of AIDS Research, a project in buildings and facilities appropriation, and the National Library of Medicine were earmarked, totaling $1.4 billion, or 6.8% of the total NIH budget (US Congress, 2002b). Within the entire Department of Health and Human Services (DHHS), $142 million, or 0.6% of the total research budget, is allocated to research performed at congressional direction (OMB, 2002a).

---

[2]The FY 2002 $9.2 million in food program and evaluation funds at ERS and the $25.5 million at NASS for the Agricultural Census represent the transfer of a program from one agency (the Food and Nutrition Service and the Commerce Department, respectively) to another.

Although CSREES administers a category of earmarks called "special grants programs" the agency has no power over the choice or amount of funding; the funds are awarded on the basis of political priorities rather than through an external peer-review process or a legislated formula. Although recent legislation (US Congress, 1998) reduced the length of special grant awards from 5 years to 3 years, repeated appropriations can still occur.

Although the outcome of the funding process for such grants is beyond the control of REE agencies, they do have complete control over the process for awarding such grants. Many federal agencies, such as the Environmental Protection Agency (EPA), require a proposal from the recipient and subjecting it to external peer review. If scientific deficiencies are identified, agencies may insist that these be addressed before the grant is awarded. CSREES reported to the committee in its telephone interviews that it has also used peer-review mechanisms to improve the quality of science in special grant proposals.

Shifts in national priorities can and should cause dramatic changes in the focus of food and agricultural research. Examples of special initiatives include some with a wide-ranging focus, such as the National Food Safety Initiative created by executive order in 2000, and some that are more narrowly focused, such as those addressing biobased products and bioenergy. The FY 2003 budget request for ARS, for example, included proposed increases in funding of several initiatives: emerging, re-emerging, and exotic diseases of animals ($8 million), biosecurity ($5 million), emerging and exotic diseases of plants ($5.4 million), and new uses for agricultural products ($9 million) (USDA, 2001b).

Special initiatives may originate in the administration or in Congress; they may respond to either broad or narrow (for example, commodity-focus) concerns or constituencies. Special initiatives may not necessarily be accompanied by additional resources. In most cases, they require reallocation of human and financial resources in the REE agencies or research institutions and potential disruptions in other important research programs. Some of the more narrowly focused national initiatives are similar to the special grants awarded by CSREES that also bypass normal formula-based and competitive funding mechanisms. Unlike special grants, however, a national initiative may be intended for one particular research institution.

**FINDING: Earmarks and national initiatives reflect the needs of particular stakeholders as articulated through the political process. Quality-assurance mechanisms, such as peer review, offer a way for REE agencies to improve the scientific quality of a special grant once it is awarded.**

## REE AND AGENCY DECISION-MAKING

A combination of REE and agency strategic planning, congressionally mandated funding mechanisms, and discretionary decisions by the REE agencies and state recipients of formula funds determines the allocation of appropriated funds to various research needs. A relatively small proportion of USDA resources is available for flexible use to address new and emerging research needs.

## Strategic Planning

The last 2 decades of the 20th century brought strategic planning and priority-setting to industrial and marketing firms. The GPRA extended strategic planning, priority-setting, and accountability to all federal agencies (US Congress, 1993). The GPRA requires strategic planning and annual program-performance reporting by every agency of the federal government, including the REE agencies. The legislation was stimulated by the perceived needs in Congress for greater accountability to taxpayers for the performance of federal programs and for better planning of federal programs. More strategic planning has been accomplished, but the committee concluded that the alignment among the agencies' individual strategic plans and the plans' connection to agency missions and actions are uneven at best, as is the agencies' implementation of effective performance measures.

Looking across the REE agencies yields a mixed picture of how well the agencies are positioned to adopt this report's vision for agricultural research (Chapter 1) into their strategic planning, according to the committee's analysis of the agencies' 5-year strategic and performance plans, lists of "future challenges" submitted to the committee, and identification of recent program accomplishments (USDA, 1997, 1999a, 2000a, 2000c, 2000e, 2000f, 2001a, 2001c).

**RECOMMENDATION 2: The REE agencies need to identify clearly their unique positions relative to the other components of the agricultural-research system, identify high-impact activities through which targeted funding and resources could generate substantial and measurable progress toward meeting national needs, and coordinate planning and research support across the agencies to minimize unnecessary duplication and maximize effectiveness. Those efforts should be informed by a clear articulation of the major national priorities for research and education and a system for anticipating, reporting on, and identifying strategies to address emerging research needs.**

Neither coordination with other research institutions nor strategic positioning currently appears to play an important role in the REE agencies' short- or long-term planning. For example, there is little evidence that the agencies explicitly set priorities according to where their research investments might play a unique or critical role or yield the greatest impacts in advancing national goals. Similarly, there appear to be no mechanisms for reviewing the research portfolios of the various REE agencies in specific topics to evaluate their combined ability to make progress toward meeting national needs. Instead, with few exceptions, current coordination appears to be largely piecemeal and ad hoc. Among the stated goals of the REE agencies, the committee was unable to identify with any clarity the top few concise research goals that they are collectively seeking to meet.

In response to GPRA and other directives under the 1998 AREERA (US Congress, 1998), each REE agency appears to have developed its strategic plan independently, in spite of the frequent meetings of the REE undersecretary and senior members of the four agencies to ensure that the strategic plans would conform to the GPRA and AREERA processes and meet USDA criteria. Some REE agency plans seem to lack alignment with the larger USDA goals and plans; a clear example is the mismatch between projected work of NASS with the overarching environmental goals and objectives in USDA's strategic plan (USDA, 2000f, 2000h, 2001d).

**FINDING: The committee finds that REE priorities would be strengthened if planning activities were more integrated, aligned, and collaborative among the agencies. It is difficult to evaluate the agencies' collective progress toward accomplishment of major national goals.**

### Allocation of Resources to Strategic Goals

Two important consequences of misalignment among agency plans and lack of focus are the inherent difficulty of accurately tracking research funding vis-à-vis today's goals and the potential difficulty of tracking funding in the future, especially as goals expand and diversify. For example, the REE agencies adopted five strategic goals that were developed in 1996 and are loosely connected to four strategic USDA goals (USDA, 2000h; 2002b); in contrast, ARS tracks expenditures for 22 national program areas, and the Current Research Information System (CRIS) tracks activities on the basis of research problem areas. Comparison of research across agencies is virtually impossible given the lack of standard definitions of research categories and nonuniform tracking methods.

Figure 4-3 (Appendix Table F-6) shows the distribution of REE funds for the five REE strategic research goals. In the committee's view, the strategic plan does not seem to guide resource allocation. About half the current REE resources is devoted to traditional agricultural productivity (goals 1 and 2), and the other half supports programs in human health, environment, and rural communities. Similarly, the CRIS data shown in Appendix Table F-7 demonstrate that about half the REE funds supports traditional agricultural productivity and enhancement, 15% human health, 15% the environment, and smaller amounts to food processing and socioeconomic research.

**FINDING: REE invests substantially in the broad array of research goals related to agriculture, food, health, environment, and communi-**

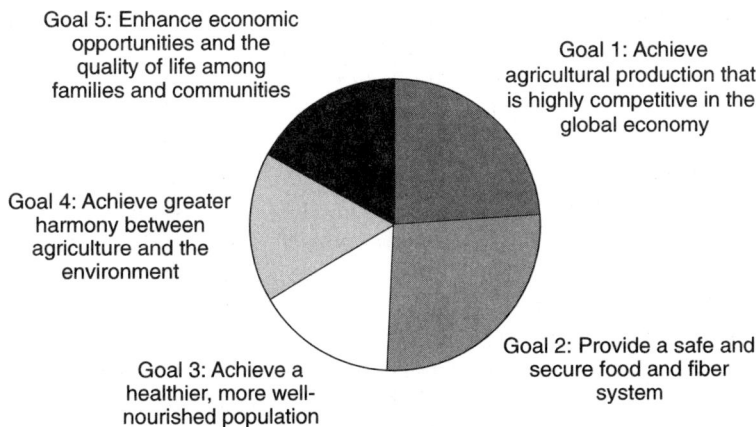

**FIGURE 4-3** FY 2000 funding allocation by REE goal (also see Appendix Table F-6). Source: Agency FY 2001 Performance Plans (USDA 2000a, 2000c, 2001a, 2001c).

ties. However, research in agricultural productivity still has the dominant share of research resources, particularly in intramural research. Furthermore, within each of the broad goals it is difficult to determine, given the limitations of existing investment tracking mechanisms, whether REE is addressing the most important opportunities.

**RECOMMENDATION 3: The REE agencies should direct new and existing resources that currently support agricultural productivity research toward new research opportunities in health, environment, and communities.** (Research opportunities are identified in Chapter 3.)

### Directing Research Toward USDA Action Agencies' Needs

The USDA action agencies[3] constitute a special group of clients for REE research because of their shared obligation to advance USDA's overall mission and goals. With USDA's reorganization in 1994, research functions for regula-

---

[3]USDA action agencies administer government programs mandated through the department. USDA action agencies are the Agricultural Marketing Service; the Animal and Plant Health Inspection Service; the Center for Nutrition Policy and Promotion; the Farm Service Agency; the Food and Nutrition Service; the Food Safety Inspection Service; the Foreign Agricultural Service; the Forest Service; the Grain Inspection, Packers, and Stockyards Administration; the Natural Resources Conservation Service; the Office of Community Development; the Risk Management Agency; the Rural Business Cooperative Service; the Rural Housing Service; and the Rural Utilities Service.

tory and action agencies were placed in REE to ensure scientific objectivity. The committee interviewed senior administrators of the action agencies about their interactions with REE agencies; the quality, timeliness, and usefulness of the research products delivered; their views of the capacity of REE agencies to meet future needs; and suggested improvements (see Appendix E).

Action agencies described a number of formal and informal processes for communicating their research needs to the REE agencies, including informal scientist-scientist interactions, ad hoc arrangements, formal memoranda of understanding, annual meetings, such bottom-up processes as the Partnership Management Agreement,[4] colocation of facilities with ARS laboratories or land-grant university campuses, involvement of action-agency researchers in research projects, and hiring of a permanent agricultural-research coordinator to interact with the REE agencies. Informal mechanisms were considered effective, but many agencies were in the process of formalizing the mechanisms for interaction, and some felt that more formal processes would be better. Colocation of staff on ARS facilities and hiring of a full-time staff member as liaison were cited as mechanisms with substantial benefits. Suggested improvements included greater discretion in the REE agencies to target money toward high-priority issues, better coordination and communication (in both directions), and greater engagement of action agencies in REE requests for proposals (RFPs) and stakeholder sessions. Responsiveness and timeliness were cited as subject to improvement, although action agencies did provide examples of REE responsiveness to data and research needs.

Many agency administrators described a divergence between REE research and action-agency needs. The quality of research was generally considered excellent but sometimes not usable or not aligned with research needs of the action agencies. Delivered research products were sometimes too complicated to be used easily or were not tailored to the action agency's needs. Some agencies also noted that REE agency staff did not always have the right mix of skills to help them. Several agency administrators cited examples in which REE had addressed short-term needs effectively, but all the agency representatives interviewed expressed concern that REE was weak in addressing their longer-term and emerging needs, funding follow-up research, and conducting applied research, systems-level integrated research, and research on programmatic, policy, or accountability questions, such as program redesign. Lack of up-to-date data was cited as a concern by one agency. The absence of incentives in REE for doing such research, which tends not to result in peer-reviewed publications, was cited

---

[4]The Partnership Management Agreement is a 3-year-old memorandum of understanding between CSREES, ARS, and NRCS. It is a process to extract and set priorities for research needs from the Natural Resource Conservation Service's (NRCS) 3,000 field offices, which are transmitted to the research community. The Partnership Management team comprises representatives of ARS, CSREES, and NRCS, with disciplinary expertise in engineering, biology, resource economics, and so on.

as a possible reason for these deficiencies, and restructuring the REE rewards system to provide incentives for doing applied or integrated research for the action agencies was proposed as a possible improvement. Action agencies noted that they lacked formal mechanisms for assessing the quality of REE support. Developing formal monitoring capability and increasing action-agency expertise to evaluate research products were cited as possible improvements. Several action-agency representatives spoke of the need for greater flexibility to seek expertise from federal, academic, or private institutions and noted that they have sought or would like to seek expertise from a variety of other institutions: consultants, private universities, and other federal agencies. Some considered these institutions capable of addressing applied-research needs more quickly or better than the REE agencies. For example, traditional agricultural colleges and ARS are not equipped with technology and expertise in such fields as x-ray technology and noninvasive monitoring.

The committee believes that allocating discretionary resources to action agencies for research could contribute to meeting action-agency research needs more effectively. REE would be well placed to receive these resources, but a more competitive mechanism would create greater accountability and transparency in terms of carrying out research designed to meet the needs of action agencies. One caution in considering this option is that provision of research services by REE helps to keep action agencies honest by providing answers to research questions that may or may not be the desired answer. Scientific expertise in the action agencies or external reviewers should be called on for help in evaluating the scientific merit of research conducted by multiple players, thereby ensuring that action agencies do not simply contract out for the answer that meets their needs. An additional caution is that the provision of discretionary funding of action agencies should in no way interfere with REE's mandate to conduct nonremunerated research that serves action agencies.

**FINDING: REE has a mixed track record in meeting needs of action agencies. More-effective mechanisms are needed for directing research toward the action agencies' long-term and emerging needs.**

## Research Funding Mechanisms

Mechanisms established by each REE agency's authorizing and appropriations legislation determine the processes by which funds, capacity, and resources are allocated to various research needs. Primarily four mechanisms are involved: formula funds, peer-reviewed grants, special research grants, and intramural funds (Figure 4-4; Appendix Tables F-8a and F-8b). Each makes a unique contribution to the fabric of agricultural research, and the diverse portfolio of approaches reduces risk. Because of differences in funding mechanism, levels of flexibility and discretionary decision-making vary substantially among the REE agencies.

Overall, a relatively small proportion of REE funds is available for flexible targeting of new research needs.

**Formula Funding**

Formula funding, based on a formula related to rural and farm populations,[5] is distributed directly to SAESs. Decision-making about how these funds are used occurs at the state level rather than within CSREES. The historical rationale for formula funding is that it ensures the pursuit of agricultural research across all states whose economies rely on agricultural production and rural livelihoods (NRC, 1996).

Formula funds, which make up just under half of research funds administered by CSREES (Figure 4-4; Appendix Table F-8a and F-8b), have a number of advantages and disadvantages. On the one hand, formula funds provide sustained support for building capacity, for long-term needs (including maintenance research), and for assisting the experiment stations in addressing research problems peculiar to their states. Formula funds tend to promote multidisciplinary research (NRC, 1996). Formula funding has also contributed to the division of responsibilities of many faculty appointments among some combination of research, extension, and teaching, which promotes linkage among research, extension, and teaching (NRC, 1996). Formula funding has minimal transaction costs and reduces the proportion of researchers' time spent applying for competitive grants (Huffman and Just, 1999). Finally, formula funding amplifies resources for agricultural research by requiring matching funds from state governments. Accountability of the formula funding system has recently improved; institutions now need to document a peer-review process to receive formula funds under AREERA (see Chapter 6 for discussion) (US Congress, 1998). On the other hand, the formula funding system lacks a mechanism for matching researchers who are uniquely qualified to address problems with a particular research question (Alston and Pardey, 1996).

On the issue of the formula itself, the committee acknowledges that the current formula no longer reflects current conditions, given that the inhabitants of rural areas no longer represent the agricultural population and that only a very small percentage of the US population works in actual production systems. However, the committee maintains that changing the formula would be politically impractical and unpopular and would pose substantial transaction costs. Any formula, including this one, develops a political interest-group following, and one can only imagine the large political debate that would ensue over a change in

---

[5]The formula, as articulated in the revised Hatch Act of 1955, is 20% of the pool of funds allocated equally among all 50 states, 26% allocated according to a state's share of the national farm population, 26% according to a state's share of the national rural population, 25% to regional, now multistate, research, and the remainder to administration (US Congress, 1887).

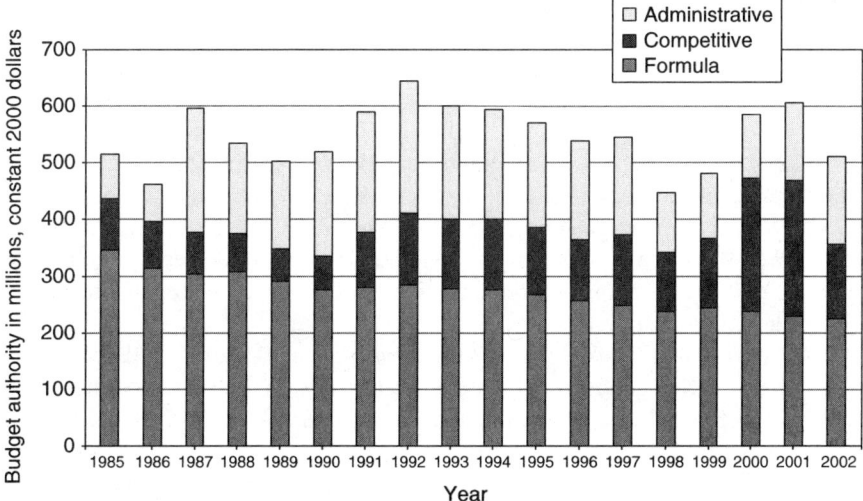

**FIGURE 4-4** Total CSREES research funding by function, constant 2000 dollars. Administrative funds include earmarks, or "special grants." Source: USDA, Office of Budget and Program Analysis (2002a).

the formula. The committee's view is that the current formula carries out its objective with little transaction cost and no political lobbying, delivering a large share of the appropriated funds to the states. Hence, in our judgment, the social benefits of changing the formula do not outweigh the costs; keeping the current formula has strong appeal. It is interesting to note, in addition, a recommendation by the National Research Council Committee on the Colleges of Agriculture at the Land Grant Universities that a new formula be designed and implemented to reflect more accurately the full range of food and agricultural research beneficiaries (NRC, 1996). It should not surprise us that such a recommendation has not been implemented, given its weak political support.

### Competitive, Peer-Reviewed Grants

Peer review has traditionally been an important mechanism for judging the quality of research outputs. Given that research quality is difficult for nonexperts to judge, judging scholarly work by other scholars with the expertise needed to assess its merits is necessary for verifying claims of discovery and for hastening the disclosure of discoveries. Peer review has been a critical foundation for open communication of scientific results, setting a precedent in science that the first to publish a discovery, idea, or finding obtains the credit.

External peer review of *prospective* research (that is, as described in research project proposals) is a relatively recent phenomenon for USDA. USDA initiated its first competitive-grant program in 1977. NSF and NIH have used external peer review for guiding the allocation of research funds since the 1950s (US GAO, 1994).

Competitive, peer-reviewed grants for several research and education programs make up about one-fourth of the research funds awarded and administered by CSREES in 2002 (Appendix Tables F-8a and F-8b). National Research Initiative (NRI), IFAFS, and FRA grants constitute the largest competitive-grant programs. Box 4-1 contains a comprehensive list of other CSREES-administered competitive-grant programs.

In general, competitive grants have been an exceptionally small and highly variable proportion of the portfolio. In recent years, there have been increases, but these increases have not been sustained. For example, between FY 2001 and 2002, competitive programs exhibited dramatic fluctuations. In FY 2001, competitive programs constituted 14% of the total REE research budget when the IFAFS program was instituted. In FY 2002, competitive grants dropped to only 9% of the REE research budget.[6] USDA (8.5%) is far below average in the R&D support allocated via merit-reviewed competitive processes[7] relative to other federal research programs, which in 2002 allocate 27% (EPA), 15% (DOE), 19% (Department of Commerce), 49% (NASA), 52% (Department of Veterans Affairs), 100% (Department of Education), 85% (DHHS), and 89% (NSF). Federal agencies allocating less R&D support with merit-reviewed competitive programs include the Department of Defense at 5.2%, the Department of the Interior at 0.5%, the Department of Transportation at 0%, and the Smithsonian Institution at 0% (OMB, 2002a).

Competitive, peer-reviewed programs have a number of advantages and disadvantages. Peer-reviewed projects, in contrast with formula funds, are considered to be more flexible and can be more responsive to newly emerging and undersupported research topics because they can effectively engage new talent and expertise (Alston and Pardey, 1996). Merit-based, peer-reviewed competi-

---

[6]As a result of the Agriculture, Rural Development, Food and Drug Administration, and Related Agencies Appropriations Act, 2002 (US Congress, 2002a), CSREES is blocked from administering a FY 2002 competition for the IFAFS. CSREES did not administer a FY 2002 competition for FRA.

[7]"Merit-reviewed research with competitive selection and external (peer) evaluation" is defined by the Office of Management and Budget as "intramural and extramural research programs where funded activities are competitively awarded following review by a set of external scientific or technical reviewers (often called peers) for merit. The review is conducted by appropriately qualified scientists, engineers, or other technically qualified individuals who are apart from the people or groups making the award decisions, and serves to inform the program manager or other qualified individual who makes the improvement of prototypes and new processes to meet specific requirements." This category is distinct from "merit-reviewed research with limited competitive selection" and "merit-reviewed research with competitive selection and internal (program) evaluation" (OMB, 2002b).

## BOX 4-1
### CSREES Competitive-Grant Programs

1890 Institution Teaching and Research Capacity Building Grants Program
Addressing Food Quality Protection Act Issues
AgrAbility Project
Agriculture Risk Management Education Competitive Grants Program
Agricultural Telecommunications Funding Program
Alaska Native-Serving and Native Hawaiian-Serving Institutions Education Grants Program
Biotechnology Risk Assessment Research Grants Program
Community Food Projects Competitive Grants Program (CFPCGP) Pest Management Alternatives Research: Special Program
Community Supported Agriculture (CSA)
Environment: SUNEI: SAES/USDA-CSREES National Environmental Initiative Research and Grants Opportunities
Food and Agricultural Sciences National Needs Graduate Fellowships Grants Program
Fund for Rural America
Higher Education Challenge - (HEC) Grants Program
Higher Education Multicultural Scholars Program
Hispanic-Serving Institutions Education Grants Program
Initiative for Future Agriculture and Food Systems
Integrated Pest Management (IPM) Program
Microbial Genome Sequencing Project
The National Integrated Food Safety Initiative
National Research Initiative Competitive Grants Program
Nutrient Science for Improved Watershed Management Program
Secondary and Two-Year Postsecondary Agriculture Education Challenge Grants Program
Small Business Innovation Research (SBIR) Program
Special Research Grants Program, Citrus Tristeza Research (CTV)
Special Research Grants Program Potato Research
Sustainable Agriculture Research and Education Program
Tribal Colleges Education Equity Grants Program (TEE)
Tribal Colleges Extension Program
Tribal Colleges Research Grants Program (TCR)
Youth Farm Safety Education and Certification Program

tive funding generally ensures that funding goes to the best proposal for a particular targeted topic (Chubin, 1994).

Critics of competitive, peer-reviewed programs cite a number of disadvantages, including lack of suitability for conducting long-term research, the time-consuming nature of proposal preparation and review (Huffman and Just, 1999; US GAO, 1994), the relatively short duration of grants, the low success rate of applications (NRC, 2000), and the long period between submission of an application and notification of award. The high transaction costs of peer-reviewed proposals do not necessarily lead to better research outcomes (Huffman and Just, 2000), and ex ante proposals do not always guarantee the quality of research. Critics also cite a tendency for risk-averse projects or scientists to be chosen (US GAO, 1994).

Alternative mechanisms for allocating competitive grants—including the use of preproposals, increasing the size and duration of grant awards, funding proposals for only a percentage of the amount requested or funding by merit percentile, reducing the time between proposal submission and award decisions, establishing continuous funding cycles for the granting process (as is the case at NSF), and performance-based funding—could be used by REE to circumvent some of these disadvantages and transaction costs (NRC, 1996). Programs to counter the risk-averse tendency of grant programs could be institutionalized. For example, at NIH, the Shannon competitive awards were developed for "enabling applicants to test the feasibility of innovative approaches, developing further tests and refining research techniques, performing secondary analyses of available data sets, and conducting discrete projects that can demonstrate research capabilities or lend additional weight to an already meritorious application" (NIH, 2002).

CSREES reported to the committee progress in several of those areas. In FY 2000, the average NRI award size was $180,473 for 2.4 years. In FY 2001, it was $188,116 for 2.4 years, approaching the targets recommended by a National Research Council study panel of average grant awards of $100,000/year and funding periods of 3 years (NRC, 2000). The success rate for NRI awards is also increasing; CSREES reported that in FY 2001, the success rate was 23.4%, an increase of 3.3% from that of FY 2000.

**Special Research Grants**

Special research grants are based on 1965 legislation (US Congress, 1965) that funds grants for selected projects. They provide the opportunity for Congress to target money for specific research topics and are an additional means of supporting land-grant experiment stations in working with federal agencies on issues of national concern. Special research grants often arise from special-interest lobbying and thus provide an additional mechanism of response to concerns of stakeholders, particularly those with good political connections.

Special research grants were the funding source for the 1890 institutions in a continuing effort to integrate them into the agricultural research system. Now the schools are funded by a USDA line item. These grants also support regional programs such as centers of excellence and rural development centers. For example, during the energy crisis in the 1970s, special grants were provided for researchers to develop ways to reduce dependence on petroleum-derived fuels. Because special grants are often determined by congressional mandate, the science supported by them might not be of the best quality available (see Chapter 6), and they might not always be directed toward the highest-priority national needs (see discussion earlier in this chapter).

**Intramural Funds**

REE has a substantial intramural research program in ARS and ERS, funded at $1.3 billion in FY 2002. ARS supports a blend of basic and applied research, including both long-term and high-risk activities. ARS has the largest component of the REE research portfolio. ARS research emphasizes agricultural problems that are national or regional in scope and related to the interests of the nation as a whole. ERS conducts intramural research that supports the needs of decision-makers in the food and agricultural sector, often in response to the needs of USDA policy-makers. Both agencies carry out unique research programs that are not duplicated in the SAESs or the private sector.

Given that research is labor intensive, a majority of intramural research resources is devoted to scientist salaries, thereby limiting the agencies' flexibility in dealing with new or emerging research topics. USDA's combined intramural research program—including REE, the US Forest Service, and other agencies—is at 72% of research funds, large in comparison with those of 8 of the 10 other federal departments or agencies tracked by NSF (NSF, 2002a). The only other federal departments that allocate a large percentage of R&D funds to support intramural research are the Department of the Interior, which allocates 88% to intramural research, and the Department of Commerce, which allocates 70%. In other mission-driven agencies—such as DHHS (which includes NIH), DOE, and EPA—the range of intramural funding is 2 to 33%.

Cooperative agreements are a funding mechanism used by the intramural research agencies that require collaboration or joint efforts by cooperating organizations to achieve a common goal or objective. They provide some flexibility in addressing new research topics, drawing on expertise outside REE, and strengthening collaboration between REE scientists and other scientists in the public sector. Cooperative agreements and other extramural agreements accounted for 13% of ARS budget authority in 2001 (Appendix Table F-9). Extramural activities (contracts, research agreements, and grants) accounted for 14% of the total ERS appropriation in FY 2001. The committee believes that such agreements should be increased to achieve greater flexibility in addressing new research topics, and that an increase to 25% of each agency's portfolio would be appropriate. Competitive mechanisms should be used for awarding large cooperative agreements ($1 million or more). However, competitive mechanisms need not be used in the case of small awards, for example, those used in hiring specialized skills in which staff are deficient or in acquiring specific skills that could best be provided by a particular scholar. For such funds, competitive processes may incur delays and higher transaction costs. In the past, REE's intramural agencies have used both competitively and noncompetitively awarded cooperative agreements. ERS has awarded cooperative agreements both competitively and noncompetitively; in FY 2001, 75% of the funds for extramural activities at ERS were disbursed on a competitive basis. ARS's research and development contracts are competitive. ARS grants, including grants for less than $75,000 (the vast majority of the ARS grants for such things as support of conferences), may or may not be competitive. ARS cooperative agreements and research-support agreements allow noncompetitive awards.

**FINDING: The diversity of financial sources usually ensures that local, state, regional, and national agricultural research needs are addressed, and some economic evidence suggests that the diversity has been a historical strength of the USDA research system. It is unclear whether the current portfolio of funding mechanisms will adequately address the complex problems of contemporary agriculture in the 21st century and realize the new vision of REE research.**

The dramatic changes in science and technology, globalization, emerging needs, and the identification of new research themes commensurate with a broader scope of societal issues also will demand greater flexibility, discretion, and collaboration in how funds are deployed by the REE agencies. A variety of mechanisms could be used to achieve these outcomes. For example, the committee believes that a realignment of the existing research budget to increase the proportion of funds in competitive grants and cooperative agreements would be effective in achieving greater flexibility and for addressing new and emerging issues by engaging new talent and expertise. Greater discretion to move resources to new areas could be achieved through no-year-funding or revolving-funding authority, or by withholding a percentage of discretionary funds for research in new areas. Discretionary funds withheld above the agency level could be used as an incentive for agency collaboration on emerging issues or emergency needs.

**RECOMMENDATION 4: To ensure that research funds are used to advance science in new directions and to address emerging and emergency issues in a timely and responsible fashion, the committee recommends the following:**

1. Total competitive grants should be substantially increased to and sustained at 20–30% of the total portfolio.
2. Action agencies should receive or control discretionary funds to be used to meet critical programmatic needs complementary to those currently served by REE agencies. The agencies could thereby fund intramural USDA scientists, other agency scientists, or university researchers competitively on the basis of the researchers' availability and match of expertise to agency needs.
3. The REE agencies should pursue complementary research activities and tap broader expertise by dedicating a higher percentage of new funds to cooperative arrangements, to be awarded on a competitive basis for large awards, with academic or other public-sector researchers.
4. Congress should increase REE budgetary flexibility to move resources toward emerging and emergency needs.

### Ensuring Relevance and Informing Decision-Making Through Stakeholder Input

AREERA (US Congress, 1998) requires USDA to consider recommendations from people who conduct or use agricultural research, extension, or education in setting department and agency priorities. The legislation has resulted in numerous changes, many just starting to be implemented, that have fundamentally altered how REE sets priorities. The most substantial change is that each REE agency now invites input from a wide variety of stakeholders into its research activities.

### Who Are the Stakeholders?

This report uses the word stakeholder in its broadest sense to mean any person or group that uses or is affected by the research, extension, and education activities conducted by REE. The term captures the people and organizations typically identified as the customers, clients, or constituents of agricultural research.

Historically, the most visible stakeholders of agricultural research have been producers, processors, and commodity groups. However, the new vision set forth in this report requires a new definition of the stakeholders of agricultural research to reflect the broadening role of agriculture in public health and nutrition, environmental stewardship, and the social and economic well-being of rural communities. Examples of the new stakeholders are producers of pharmaceutical products; sustainable-, alternative-, and organic-farming interests; a broad array of public and private natural-resource and land managers; conservationists; and rural communities and government agencies. The increased breadth is bringing new ideas and insights into the REE research endeavor. It also poses the challenge of combining diverse stakeholder concerns to help shape a cohesive and feasible research program.

**Mechanisms of Stakeholder Input**

Mechanisms for integrating stakeholder input into the research process range from formal, national, advisory boards to cooperative extension county-level meetings and informal working relationships between scientists and users of research findings (summarized in Table 4-1). Although the REE agencies have developed a wealth of information through various forms of stakeholder input over the last several years, the overall experience has been mixed. Important issues have arisen about how to ensure balanced input and how to translate the

**TABLE 4-1** REE Mechanisms for Ensuring Stakeholder[a] Input

| Mechanisms | Agency Using Mechanisms |
| --- | --- |
| National Agricultural Research, Extension, Education, and Economics Advisory Board | ARS, CSREES, ERS, NASS |
| Agency-specific advisory boards | NASS, ERS |
| Public workshops and listening sessions | CSREES, ARS |
| Stakeholder input at the state level (through field offices and universities) | NASS, CSREES |
| Stakeholder input in competitive-grant RFPs | CSREES |
| Stakeholder participation in research and extension projects | CSREES |
| Informal or ad hoc communication of priorities between REE agencies and USDA regulatory and action agencies | ARS, CSREES, ERS, NASS |
| Formal partnership agreements between REE agencies and USDA regulatory and action agencies | ARS, CSREES |
| Formal annual meetings between REE agencies and USDA regulatory and action agencies | ARS |
| Colocation of action agency staff on REE facilities; involvement of action-agency staff in research | ARS, CSREES |
| Action-agency staff full-time liaisons at REE agencies | ARS |
| Communication of priorities through state departments of agriculture and commodity groups | CSREES, ARS |
| Input through respondent interviewers | NASS |

[a]Includes action-agency or regulatory-agency input.
Source: Data provided to the committee by REE and action agencies in 2001 and 2002.

frequently overwhelming amounts of information and diverse perspectives into focused research priorities.

The most important advisory board for the REE agencies is the National Agricultural Research, Extension, Education, and Economics (NAREEE) Advisory Board, which was established by the 1996 farm bill (US Congress, 1996) and draws members from 30 constituencies identified by the legislation (Table 4-2; Lechtenberg, 2001a). Its role is to provide overall guidance to the REE mission area on policies and priorities for agricultural research, extension, education, and economics. The board sponsors stakeholder listening sessions, reviews

**TABLE 4-2** Membership Categories, Represented in NAREEE Advisory Board

| Membership Categories |
| --- |
| National farm organization |
| Farm cooperative |
| Food-animal commodity producer |
| Plant-commodity producer |
| National animal-commodity organization |
| National crop-commodity organization |
| National aquaculture association |
| National food-animal science society |
| National crop, soil, agronomy, horticulture, or weed science society |
| National food-science organization |
| National human-health association |
| National nutritional-science society |
| 1862 land-grant college |
| 1890 land-grant college |
| 1994 institution |
| Hispanic-serving institution |
| American college of veterinary medicine |
| Nonagriculture scientific community |
| Food and agricultural products transporter, for both domestic and foreign markets |
| Food retailing and marketing representative |
| Food and fiber processor |
| Rural economic development advocate |
| National consumer interest group |
| National forestry group |
| National conservation or natural-resource group |
| Private-sector international development organization |
| USDA nonresearch agency |
| Non-USDA federal government research agency |
| National social-science association |
| National agricultural research, education, and extension organization |

Source: Federal Agricultural Improvement and Reform Act, US Congress (1996).

draft guidance for competitive-grant programs, and conducts annual reviews of the REE portfolio for relevance and adequacy of funding.

In addition to this overall advisory board for the REE agencies, both NASS and ERS have specific advisory groups that address programmatic areas unique to each agency. A 25-member Advisory Committee on Agriculture Statistics advises NASS on the scope, content, and timing of the agricultural census and related surveys. ERS has convened "roundtables," which include commodity and trade association representatives, to gain feedback on commodity-related issues and other aspects of ERS's market analysis and outlook program. A group of scholars, researchers, and policy officials also reviews ERS research priorities for the Food and Nutrition Assistance Research Program and provides guidance on its scope and direction.

Those boards sometimes provide valuable input to REE agencies, but questions have arisen about their ability to address the breadth of the future REE research portfolio. For example, the NAREEE Advisory Board has provided recommendations to the secretary of agriculture on research to serve small farms (USDA, 2000g) but has rarely commented on environmental stewardship, despite identifying it as a high-priority item (Lechtenberg, 1998, 2001b; Lechtenberg and Dooley, 1999, 2000). Similarly, the membership of NASS's advisory committee matches the agency's *current* heavy emphasis on production agriculture but does not reflect other important components of the food system, such as processing, manufacturing, distribution, retailing, consumption, waste management, natural resources, and the environment.

Agency administrators reported to the committee in interviews that open workshops or listening sessions conducted by REE agencies across the country have provided an opportunity for many more stakeholders to present information and their perspectives. For example, NASS's listening sessions have taken the form of data user meetings held in cooperation with partnering agencies. A recent environmental-data users meeting held by NASS and ERS made several important recommendations—such as to increase integrated pest management data collection, to collect socioeconomic data, and to survey seed use (USDA, 2000h). The committee observed that the listening-session approach used by all the agencies tends to be weighted toward stakeholders who have the time, money, or desire to participate in the meetings. Those stakeholders are often well-funded industry groups rather than less well-funded stakeholders, such as small farmers and environmental organizations.

Overall, the REE agencies appear to have an inconsistent track record of effectively using information generated in public workshops. ARS, for example, has developed programs (for example, related to food safety, small farms, and organic farming) in response to stakeholder input but in the committee's analysis has not always made full use of material from its national program planning workshop summaries (USDA, 1999b, 2000b) in developing its national program action plan.

CSREES meets the requirements of AREERA (US Congress, 1998) in formula-funding programs by requiring state institutions to report how they gathered stakeholder input and to submit a plan of work. CSREES reported to the committee that states use various methods to gather input, such as dean's advisory boards, department advisory committees, agricultural councils, local extension boards, and random telephone surveys of citizens (e.g., University of Florida, 1999).

Stakeholders influence competitive-grants programs by providing input to RFPs and by participating in grant review and selection processes (Box 4-2). CSREES has actively solicited stakeholder comment in the development of RFPs, for example, through *Federal Register* announcements and specific requests to underrepresented constituencies. Staff members from competitive-grant programs obtain less-formal input through scientific and professional meetings, science forums, user workshops, and communication with other federal agencies, commodity and consumer organizations, trade organizations, peer-review panelists, and panel managers.

---

**BOX 4-2**
**Examples of REE Responsiveness to Stakeholder Input**

- The NRI competitive-grant program has relied on multiple forms of stakeholder input throughout its history to ensure that the research solicited through RFPs meets high-priority needs. The NRI scientific staff meets with the Animal Agriculture Coalition two or three times per year to discuss program directions and priorities; they also convened major symposia (FAIR–Food Animal Integrated Research) in 1992 and 1999 to identify needed changes in program priorities. Such interactions have led to important changes in program emphasis in several cases, such as a stronger emphasis on animal health and well-being in one program and the generation of priorities for microbial genetics research in animal agriculture.
- ERS was faced with competing pressures from stakeholders in setting priorities for its Food Assistance and Nutrition Research Program. Internal stakeholders were concerned about the potential impacts of welfare reform on able-bodied adults without dependents (a group particularly affected by rule changes). External stakeholders were concerned about the relative roles of the strong economy, rule changes, and other assistance programs in accounting for the sharp decline in caseloads during the 1990s. ERS responded by funding two dozen projects over a 3-year period that addressed both sets of concerns.

Farmers, members of nonprofit organizations, and state and local agencies are sometimes directly involved in the development and implementation of REE research projects. Such involvement can range from active leadership by farmer participants to scientists' informal consultation with collaborators or farmers. Many researchers have informal networks for their own stakeholder input. Demand for research on new topics often comes through such contacts. The Sustainable Agriculture Research and Education Program (SARE) constitutes a case study of stakeholders' involvement at several levels (Box 4-3).

## Benefits and Costs of Stakeholder Involvement

Various studies have demonstrated that stakeholder involvement can help to ensure the relevance of agricultural research, education, and development. Examples include innovative participatory research processes in which farmers collaborate with scientists and technical extension officers (Pretty, 2002; Pretty and Hine, 2001; Thrupp, 1996; Thrupp and Altieri, 2001; Uphoff, 2002; Western SARE, 2000), farmer networks and farmer-to-farmer educational methods (e.g., Flora, 2001; Pretty, 2002; Thrupp, 1996; Western SARE, 2000), and watershed management programs where local land managers and community groups help plan, implement, and evaluate related research (Thompson and Guijt, 1999). Huffman and Just (1994) showed that broad external influences on scientists, including

---

**BOX 4-3 Stakeholder Participation and SARE**

In SARE, stakeholder participation is engaged at three levels: priority-setting, project review, and project implementation. A broad group—including producers, farm consultants, university researchers and administrators, state and federal government agency staff, and representatives of nonprofit organizations—serves on the regional administrative councils that provide overall leadership for the program; establish program priorities, goals, and objectives; and select projects for funding. Stakeholders also serve on the technical boards convened by each regional administrative council to review the technical quality and relevance of SARE proposals. For example, the 2000 North Central SARE technical committee included 10 reviewers from the private sector (mostly producers) and 10 reviewers from the public sector—researchers and extension personnel from universities, ARS, the Natural Resources Conservation Service, and the US Environmental Protection Agency (USDA, 2000i). At the project level, since 1992, SARE has offered a small-grant program for farmers and ranchers to run their own on-site research experiments.

those by stakeholders, increased the impact of basic research on agricultural productivity.

Nevertheless, stakeholder involvement does entail transaction and opportunity costs resulting from the requirement of time and resources for meetings and discussions involving multiple actors in the research process and from related losses of research productivity (Uphoff, 2002). Whether those transaction costs ever outweigh the overall benefits is unclear. Experience is showing, however, that initial transaction costs often are worthwhile (Uphoff, 2002; Western SARE, 2000), particularly in cases related to environmental and natural-resources issues or other topics involving stakeholders who are new to agricultural research.

**FINDING: The REE agencies have implemented numerous mechanisms to integrate stakeholder input into their priority-setting and into the research, extension, and education processes. Stakeholder input generally strengthens the connection between research and its applications but has had mixed results. Not all processes have ensured balanced participation by the full array of affected stakeholders. Efforts have been largely unlinked across agencies, and this has created duplication of effort and sometimes disparate results. The current multitude of stakeholder processes taxes stakeholder time and resources and the already-stretched capacity and resources of the REE agencies. Moreover, the agencies have sometimes found it difficult to reconcile stakeholders' competing views and to synthesize diverse and abundant stakeholder input into a usable form. Finally, stakeholder processes are weakly linked to REE and the agencies' strategic-planning and performance-evaluation processes.**

**RECOMMENDATION 5: To provide a forum for shared learning across agencies, REE should conduct a national summit every 2–3 years that would engage the four REE agencies and a broad representation of stakeholders at the local, national, and regional levels. The summit could assess national research needs and inform stakeholders how their input is used in agency decision-making.**

Such a national summit could include a preliminary series of open workshops conducted in collaboration by all the REE agencies at local, state, and regional levels. Those meetings would utilize the national network of cooperative extension and other mechanisms at the state, local, and regional levels to develop information on research needs. It could affect research decision-making at all levels. Results from the summit could be integrated into REE's and its agencies' strategic plans, performance assessments, and decision-making. Web-based communication could also be used to solicit input from stakeholders and to disseminate summit results to stakeholders and to the broader research community.

Confusion often exists among stakeholders regarding the mission and responsibilities of various research agencies in REE. This confusion is reinforced by some of the tensions among the agencies that are not necessarily overt but exist nonetheless. The concept of a national summit will have greatest value if it is well coordinated among the agencies and not used to reinforce existing tensions. In a true visionary mode, such a summit should be used to articulate the need for coordination and collaboration among the mission areas of REE.

## SUMMARY

This chapter has described federal resource allocation to agricultural research in the context of total federal R&D funding, in the context of state and private funding for agricultural research, and in comparison with agricultural research funding in other countries. The proportion of federal research allocation to earmarks, special grants, and national initiatives was presented. The committee identified a need for USDA to be more strategic in the application of its limited resources.

The strategic planning process in REE and the allocation of federal resources toward various research needs based on strategic goals were analyzed, and a misalignment between strategic goals and resource allocation was noted. The effectiveness of REE agencies in serving USDA action agencies' needs was considered. Advantages and disadvantages of the four funding mechanisms used by USDA—formula funds, intramural funds, competitive funds, and special grants or earmarks—were discussed, and recommendations to realign the research budget to achieve greater flexibility and to address new and emerging issues were offered.

REE mechanisms for ensuring relevance of research to stakeholder needs were considered. Changes in the mechanisms for stakeholder input that have resulted from the 1998 AREERA were described, and the effectiveness of these mechanisms was discussed. A coordinated effort to elicit stakeholder input is recommended.

## REFERENCES

AAAS (American Association for the Advancement of Science). 2002. Historical Table 2. Total R&D by Agency, FY 1976-2003. March, 2002. Available online at *http://www.aaas.org/spp/dspp/rd/hist03p2.pdf*.

Alston, J.M., and P.G. Pardey. 1996. Making Science Pay: The Economics of Agricultural R&D Policy. Washington, DC: AEI Press.

Chubin, D. 1994. Grants peer review in theory and practice. Evaluation Review 18:20–30.

Flora, C., ed. 2001. Shifting agroecosystems and communities. Pp. 5–14 in Interactions Between Agroecosystems and Rural Communities, C. Flora, ed. Boca Raton, FL: CRC Press.

Huffman, W., and R. Evenson. 1993. Science for Agriculture: A Long-Term Perspective. Ames, IA: Iowa State University Press.

Huffman, W.E., and R.E. Just. 1994. Funding, structure, and management of public agricultural research in the United States. American Journal of Agricultural Economics 76:744–759.
Huffman, W.E., and R.E. Just. 1999. The organization of agricultural research in western developed countries. Agricultural Economics 21:1–18.
Huffman, W.E., and R.E. Just. 2000. Setting efficient incentives for agricultural research: Lessons from principal-agent theory. American Journal of Agricultural Economics 82(November):828–841.
Klotz, C., K. Fuglie, and C. Pray. 1995. Private-Sector Agricultural Research Expenditures in the United States, 1960–92. Staff Paper No. 9525. October. Washington, DC: Economic Research Service, US Department of Agriculture.
Lechtenberg, V.L. 1998. Letter to Secretary of Agriculture Dan Glickman from the National Agricultural Research, Extension, Education, and Economics Advisory Board, May 27.
Lechtenberg, V.L. 2001a. Testimony before the Senate Committee on Agriculture, Nutrition, and Forestry on behalf of the National Agricultural Research, Extension, Education, and Economics Advisory Board, March 27.
Lechtenberg, V.L. 2001b. Letter to Secretary of Agriculture Ann Veneman from the National Agricultural Research, Extension, Education, and Economics Advisory Board, January 25.
Lechtenberg, V.L. and D.M. Dooley. 1999. Letter to Secretary of Agriculture Dan Glickman from the National Agricultural Research, Extension, Education, and Economics Advisory Board, June 4.
Lechtenberg, V.L. and D.M. Dooley. 2000. Letter to Secretary of Agriculture Dan Glickman from the National Agricultural Research, Extension, Education, and Economics Advisory Board, April 14.
NIH (National Institutes of Health). 2002. James A. Shannon Director's Award Program to Continue in 1996. Available online at *http://www.csr.nih.gov/prnotes/june96.htm#shannon*.
NRC (National Research Council). 1996. Colleges of Agriculture at the Land Grant Universities: Public Service and Public Policy. Washington, DC: National Academy Press.
NRC (National Research Council). 2000. National Research Initiative: A Vital Competitive Grants Program in Food, Fiber, and Natural-Resources Research. Washington, DC: National Academy Press.
NSF (National Science Foundation). 2002a. Science and Engineering Indicators – 2002. Estimated federal obligations for R&D and R&D plant, by selected agency, performer, and character of work: Fiscal year 2001. Appendix Table 4-25. Available online at *http://www.nsf.gov/sbe/srs/seind02/append/c4/at04-25.pdf*.
NSF (National Science Foundation). 2002b. Science and Engineering Indicators – 2002. Estimated federal obligations for research, by agency and field of science and engineering: Fiscal year 2001. Appendix Table 4-27. Available online at *http://www.nsf.gov/sbe/srs/seind02/append/c4/at04-27.pdf*.
OMB (Office of Management and Budget). 2002a. Budget of the US Government, FY 2003. Analytical Perspectives. Pp. 173–174. Available online at *http://www.whitehouse.gov/omb/budget/fy2003/pdf/spec.pdf*.
OMB (Office of Management and Budget). 2002b. OMB Circular A–11. Section 84–Character Classification (Schedule C). Available online at *http://www.whitehouse.gov/omb/circulars/a11/2002/part2.pdf*.
Pardey, P.G., J. Roseboom, and B.J. Craig. 1999. Agricultural R&D investments and impacts. Pp. 31–68 in Paying for Agricultural Productivity, J.M. Alston, P.G. Pardey, and V.H. Smith, eds. Baltimore; London: Johns Hopkins University Press.
Pretty, J. 2002. Social and human capital for sustainable agriculture. In Agroecological Innovations: Increasing Food Production with Participatory Development, N. Uphoff, ed. London: Earthscan Publications.
Pretty, J., and R. Hine. 2001. Reducing Food Poverty with Sustainable Agriculture: A Summary of New Evidence, Final Report for SAFE-World Research Project. Colchester, UK: University of Essex.

Thompson, J., and I. Guijt. 1999. Sustainability indicators for analysing the impacts of participatory watershed management programmes. Pp. 13–26 in Fertile Ground: The Impacts of Participatory Watershed Management, F. Hinchcliffe, J. Thompson, J. Pretty, I. Guijt, and P. Shah, eds. London, UK: Intermediate Technology Publications.

Thrupp, L.A. 1996. New Partnerships for Sustainable Agriculture. Washington, DC: World Resources Institute.

Thrupp, L.A., and M. Altieri. 2001. Innovative models of technology generation and transfer: Lessons learned from the south. In Knowledge Generation and Technical Change: Institution Innovation in Agriculture, S. Wolf and D. Zilberman, eds. New York: Kluwer Academic Press.

University of Florida. 1999. AREERA Plan of Work. Available online at *http://pdec.ifas.ufl.edu/AREERApow.htm*.

Uphoff, N., ed. 2002. Agroecological Innovations: Increasing Food Production with Participatory Development. London: Earthscan Publications.

US Congress. 1887. P.L. (Public Law) 84-352. Hatch Act of 1887 (amended 1955).

US Congress. 1890. 7 USC. 322 et seq. 26 Stat. 417. Second Morrill Act of 1890.

US Congress. 1965. P.L. (Public Law) 89-106. Section (2), as amended (7 USC. 450i).

US Congress. 1993. P.L. (Public Law) 103-62. Government Performance Review Act.

US Congress. 1996. P.L. (Public Law) 104-127. Federal Agriculture Improvement and Reform Act (FAIR) of 1996.

US Congress. 1998. P.L. (Public Law) 105-185. Agricultural Research, Extension, and Education Reform Act (AREERA) of 1998.

US Congress. 2001a. Committee Report - House Rpt. 107–116 - Agriculture, Rural Development, Food and Drug Administration, and Related Agencies Appropriations Bill, 2002. Available online at *http://Thomas.Loc.Gov/Cgi-Bin/Cpquery/R?Cp107:Fld010:@1(Hr116)*.

US Congress. 2001b. H.R. 107-275. Committee Report—House Rpt. 107–275—Making Appropriations for Agriculture, Rural Development, Food and Drug Administration, and Related Agencies Programs for the Fiscal Year Ending September 30, 2002, and for Other Purposes. Available online at *http://Thomas.Loc.Gov/Cgi-Bin/Cpquery/R?Cp107:Fld010:@1(Hr275)*.

US Congress. 2001c. H.R. 107-41. Committee Report—Senate Rpt. 107–41—Agriculture, Rural Development, Food and Drug Administration, and Related Agencies Appropriations Bill, 2002. Available online at *http://Thomas.Loc.Gov/Cgi-Bin/Cpquery/R?Cp107:Fld010:@1(Sr041)*.

US Congress. 2002a. P.L. (Public Law) 107-76. Agriculture, Rural Development, Food and Drug Administration, and Related Agencies Appropriations Act, 2002 (Enrolled as Agreed to or Passed by Both House and Senate). Available online at *http://frwebgate.access.gpo.gov/cgi-bin/getdoc.cgi?dbname=107_cong_reports&docid=f:hr275.107.pdf*.

US Congress. 2002b. P.L. (Public Law) 107-116. Departments of Labor, Health and Human Services, and Education, and Related Agencies Appropriations Act, 2002 (Enrolled as Agreed to or Passed by Both House and Senate). Available online at *http://thomas.loc.gov/home/approp/app02.htm*.

USDA (US Department of Agriculture). 1997. CSREES Draft Strategic Plan. Washington, DC: Cooperative State Research, Education, and Extension Service, US Department of Agriculture.

USDA (US Department of Agriculture). 1999a. Agricultural Research Service Strategic Plan: Working Document 1997–2002. Washington, DC: Agricultural Research Service, US Department of Agriculture. Available online at *http://www.nps.ars.usda.gov/mgmt/stratpln/1999/background.cfm*.

USDA (US Department of Agriculture). 1999b. Animal Production Systems and Animal Health Research Program, National Program Planning Workshop Summary. Washington, DC: Agricultural Research Service, US Department of Agriculture.

USDA (US Department of Agriculture). 2000a. Agricultural Research Service FY 2000 and 2001 Annual Performance Plans. Washington, DC: Agricultural Research Service, US Department of Agriculture.

USDA (US Department of Agriculture). 2000b. Animal Genomes, Germplasm, Reproduction, and Development, National Program Planning Workshop Summaries. Washington, DC: Agricultural Research Service, US Department of Agriculture.

USDA (US Department of Agriculture). 2000c. Cooperative State Research, Education, and Extension Service FY 2000 and 2001 Annual Performance Plan. Washington, DC: Cooperative State Research, Education, and Extension Service, US Department of Agriculture.

USDA (US Department of Agriculture). 2000d. Current Research Information System (CRIS) Funding Summaries, Table A, FY 2000. Washington, DC: Cooperative State Research, Education, and Extension Service, US Department of Agriculture. Available online at http://cristel.csrees.usda.gov/star/00tablea.pdf.

USDA (US Department of Agriculture). 2000e. Economic Research Service Strategic Plan, 2000–2005. October 11. Washington, DC: Economic Research Service, US Department of Agriculture. Available online at http://www.ers.usda.gov/AboutERS/ersstrategicplan.pdf.

USDA (US Department of Agriculture). 2000f. National Agricultural Statistics Service. GPRA Strategic Plan. Washington, DC: National Agricultural Statistics Service, US Department of Agriculture. Available online at http://www.usda.gov/nass/nassinfo/strat-2005.pdf.

USDA (US Department of Agriculture). 2000g. Research and Education Recommendations for Small Farms. January 14. Washington, DC: National Agricultural Research, Extension, Education, and Economics Advisory Board, US Department of Agriculture.

USDA (US Department of Agriculture). 2000h. Summary of 2000 National Agricultural Statistics Service and Economic Research Service Environmental Data Users Meeting. October 10. Washington, DC: National Agricultural Statistics Service, US Department of Agriculture.

USDA (US Department of Agriculture). 2000i. Sustainable Agriculture Research and Education Program. North Central Regional Technical Committee. Washington, DC: Cooperative State Research, Education, and Extension Service, US Department of Agriculture. Available online at http://www.sare.org/ncrsare/leaders.htm.

USDA (US Department of Agriculture). 2001a. Economic Research Service FY 2002 Annual Performance Plan and Revised Plan for FY 2001 (July). Washington, DC: Economic Research Service, US Department of Agriculture. Available online at http://www.ers.usda.gov/AboutERS/ersperformance_plan.pdf.

USDA (US Department of Agriculture). 2001b. FY 2003 Budget Summary. Washington, DC: US Department of Agriculture. Available online at http://www.usda.gov/agency/obpa/Budget-Summary/2003/2003budsum.htm#ree.

USDA (US Department of Agriculture). 2001c. National Agricultural Statistics Service FY 2002 and Revised FY 2001 Annual Performance Plans. Washington, DC: National Agricultural Statistics Service, US Department of Agriculture. Available online at http://www.usda.gov/nass/nassinfo/nass-app-02-01.pdf.

USDA (US Department of Agriculture). 2001d. Published Estimates Database. Washington, DC: National Agricultural Statistics Service, US Department of Agriculture. Available online at http://www.nass.usda.gov:81/ipedb/.

USDA (US Department of Agriculture). 2002a. Office of Budget and Program Analysis. Data submitted to National Research Council Committee on Opportunities in Agriculture. Washington, DC: US Department of Agriculture.

USDA (US Department of Agriculture). 2002b. Research, Education, and Economics Strategic Plan. Washington, DC: US Department of Agriculture. Available online at http://www.reeusda.gov/ree/ree2.htm.

US GAO (US General Accounting Office). 1994. Peer Review: Reforms Needed to Ensure Fairness in Federal Agency Grant Selection. Report of the Chairman, Committee on Governmental Affairs, US Senate. Gaithersburg, MD: US General Accounting Office.

Western SARE (Sustainable Agriculture Research and Education). 2000. Sustainable Agriculture: Continuing to Grow, Proceedings of the Conference on Farming and Ranching for Profit, Stewardship, and Community. Portland, OR: Western Sustainable Agriculture Research and Education Program and Sustainable Northwest.

# 5

# Collaboration

A key element of the committee's vision of the future is greater collaboration to enable the US Department of Agriculture (USDA) Research, Education, and Economics (REE) mission area to address future research opportunities more effectively. Collaboration will need to be enhanced both within REE and between REE and other research institutions. This chapter considers current collaborations and mechanisms to support collaboration, including collaborative research across scientific disciplines, among agencies within REE, with other federal research agencies outside USDA, with nonprofit and international research organizations, and between research and extension. The final section of this chapter considers collaboration between the public and private sectors in agricultural research in some detail because this is a subject of growing importance.

## MULTIDISCIPLINARY RESEARCH

The success of the agriculture and food enterprise that followed the establishment of USDA and development of the Agricultural Research Service (ARS) in production agriculture through the 1980s was the result of targeted investments in meeting needs of individual states and agricultural regions. That led to production of abundant food, feed, and fiber for America and the world. To realize that success, agricultural scientists generally maintained fairly sharp disciplinary divisions in their educational background, research orientation, criteria for research-problem choice, and publication activities (Busch and Lacy, 1983; Huffman and Evenson, 1993). Disciplinary problems were likely to receive more support than research on complex applied problems that crossed disciplinary lines. Such problems are more specialized, reductionist in approach, and easier to assess in terms of disciplinary significance. However, that self-reinforcement also implied that the stock of knowledge produced by each of the disciplines could be disconnected from that of other disciplines. Moreover, by focusing on aspects of the world that are deemed relevant by a particular discipline, scientists appeared to ignore problems that resided outside their competence. Today, the increasing complexity of the issues and challenges facing our food and fiber system, the environment, and families and communities requires disciplinary, multidisciplinary, and systems-level approaches. The future success of the agriculture enterprise in solving complex applied problems will require collaborative and interactive participation across greater numbers of disciplines.

The committee observed that a key conceptual shift in the scientific foundation of agriculture has been the recognition that effective solutions to many food, health, environmental, and community-development concerns require both a strong disciplinary perspective and a multidisciplinary and integrated systems perspective. For example, research that is strictly physical and biologic will yield physical and biologic solutions, but most complex agricultural, environmental and community challenges require an equally rigorous understanding of social and economic issues. In many cases, the socioeconomic portion of a problem is as complex and unstudied as the biologic and physical and requires fundamental social-science research. For example, Matson et al. (1997) describe how social, demographic, and economic factors have affected adoption of various farming practices and therefore agriculture's impacts on ecosystem processes, and they call for research integrating social and natural sciences to develop sustainable agriculture.

A multiscale, integrated systems approach to research will yield complementary and robust scientific insight and results. It also will produce research that is more anticipatory by providing a deeper understanding of food and agricultural systems. Ultimately, effective approaches will depend on an integration of the biophysical and socioeconomic research, and the integration should occur from the outset. An integrated approach to research will enable scientists and analysts to more rapidly determine which new technologies or changing agricultural practices and policies will cause beneficial and adverse impacts and consequently provide a richer set of options for ensuring a sustainable food system, generating environmental benefits, and enhancing communities.

A systems approach to evaluating agricultural technologies, for example, requires more than understanding how effective technologies will be. In the case of new plant-based technologies, such as transgenics, a systems approach would ask such questions as, Will this particular option provide potential environmental, economic, or social benefits, either direct or indirect? For example, how might integration of the technology into existing cropping systems affect, favorably or adversely, nontarget organisms, overall biodiversity, water quality, fresh water and marine ecosystems, and community infrastructure? A systems approach that incorporates multidisciplinary research must also address economic and social viability of technologies.

Agriculture and our food system affect and are profoundly affected by human societies and behaviors. Developing policies that shape agriculture's future and serve the public good will require an understanding of societal changes—the quality of life in rural communities, aesthetics and the burgeoning land-trust movement, projected demographic and land-use changes, and the effects of globalization on local and national economies (e.g., Flora, 2001).

REE has engaged in a number of effective multidisciplinary efforts. For example, the 1990 National Water Quality Initiative provided a potential model for coordinating multidisciplinary and multilocal research and extension efforts across federal agencies to meet a national environmental-research need (Amerman et al., 2001; Caswell, 2001; Zucker and Brown, 1998). This 10-year program was a joint venture of ARS, the Cooperative State Research, Extension, and Education Service (CSREES), the Economic Research Service (ERS), the National Agricultural Statistics Service (NASS), and the Natural Resources Conservation Service (NRCS) with the Department of the Interior (DOI), the US Geological Survey, the Department of Commerce (DOC), and the Environmental Protection Agency (EPA). The aim of the program was to reduce agricultural watershed contamination by nitrogen, phosphorus, and pesticides through a combination of research, education, and outreach projects funded competitively by ARS base funds, ERS cooperative agreements, CSREES special grants, and cost-sharing and technical assistance via NRCS. The program was implemented by using five Management Systems Evaluation Areas, 149 inhouse and cooperative projects, and incentive payments for the adoption of improved farm-management systems. From its inception, the initiative used a multidisciplinary and systems-level approach, with representation of all relevant disciplines and coordinated implementation. Initially, a small working group was formed with representatives of each of the USDA agencies. Although disciplinary identity was maintained in the agencies, multiple efforts were coordinated by a steering committee at the secretary's level in USDA; this resulted in integration of the results of local research projects and outreach efforts to accomplish a national goal. This model deserves further consideration in REE agencies' strategic planning, implementation, and program execution.

Another key example within CSREES of a strong commitment to multidisciplinary approaches has been the various competitive research-grant programs, such as the National Research Initiative (NRI), the Fund for Rural America (FRA), and the Initiative for Future Agriculture and Food Systems (IFAFS). Through those programs, the agency has committed hundreds of millions of dollars to multidisciplinary work, recognizing that many of the important questions facing food, agriculture, the environment, and communities are at the disciplinary boundaries. Under legislative mandate, the NRI requires that a portion of awards be multidisciplinary, and the FRA and IFAFS have multistate and multidisciplinary requirements.

At the same time, leaders at ARS and ERS indicate that although some interaction between their staffs occurs, the interaction is not systematically organized to provide the kind of long-term multidisciplinary research needed for the future. They note that the staffs of ERS and ARS operate essentially in different domains. The ARS administrator observed in an interview with the committee that the lack of a mechanism to involve social scientists is "a major deficiency" in the national agricultural research program. In REE research programs the extent of integration, multidisciplinary research, and multidisciplinary complementarity varies widely.

Developing a systems approach will require a greater emphasis on multidisciplinary research planning and execution that combine rigorous techniques in the biologic, social, and physical sciences. Because multidisciplinary work brings together knowledge and methods from different fields, it involves fewer simplifying assumptions and can yield more robust solutions to complex problems. This kind of approach is essential for many of the new agricultural methods, processes, and technologies. USDA's research system, particularly ERS and ARS, will need to evaluate the success of multidisciplinary structures—such as task forces, centers, institutes, and initiatives—in terms of their potential application in REE.

Institutions responsible for and engaged in graduate education need to expand multidisciplinary education to include a broader understanding and appreciation of different scientific perspectives and to provide a better integration of those perspectives. Multidisciplinary interdepartmental graduate fields are promising developments, but they often lack adequate institutional support and must rely on academic departments for resources and faculty time. The restructuring of graduate education must start with policies, practices, and norms regarding curriculum, seminars, professional meetings, appropriate journals, and other key means of mentoring and professionally socializing the next generation of scientists. Such changes have the potential to strengthen multidisciplinary agricultural and food-systems research. (Examples of CSREES-funded education efforts are discussed further in Chapter 7.)

In the context of more rigorous and advanced disciplinary sciences, multidisciplinary programs risk producing "jacks of all trades but masters of none." Some multidisciplinary programs in colleges of agriculture have not always been particularly successful, especially at the PhD level. Graduate integrated pest-management programs and programs in sustainable agriculture are examples. But there are some successful multidisciplinary graduate programs, such as molecular biology programs, bioinformatics, and risk-assessment programs. Although developing multidisciplinary programs for the sake of being multidisciplinary is not useful, and not all problems need multidisciplinary approaches, new disciplines are being demanded, and the system needs to move forward to meet the demands.

## COLLABORATION WITHIN REE

Effective collaboration among the land-grant universities and their colleges of agriculture, forestry, and human ecology (CSREES) and other REE units has been in effect for a long time. With the passage of the Federal Research and Marketing Act in 1946 (US Congress, 1946), one-fourth of the formula funds (Hatch and McIntire-Stennis) were set aside for regional research, thereby stimulating many interuniversity collaborative efforts. The Agricultural Research, Extension, and Education Reform Act of 1998 (US Congress, 1998) changed the title of these programs from regional to multistate, as many of these projects are national in scope. Today, many regional research projects involve multiple states, multiple regions, multiple universities (land-grant and non-land-grant), ARS, ERS, and other profit and nonprofit organizations. Examples of REE leadership in regional research efforts are found in Box 5-1.

Regional rural development centers have been another model for effective interuniversity collaborative research in CSREES. Although they require more coordination and cooperation among scientists and more administrative support than individual-scientist projects, the regional efforts generally have been successful. Most projects have produced numerous important scientific peer-reviewed publications and policy analyses and have addressed relevant practical issues with sound science in ways not often possible through individual projects.

In addition to multistate research, ARS has successfully collaborated with land-grant universities and CSREES particularly in the plant and animal sciences. ARS has often located its research facilities close to the universities or posted its scientific staff at universities. Similarly, ERS has joined with land-grant universities in collaborative work and cooperative agreements primarily with departments of agricultural and resource economics and to a more limited extent with rural sociology, nutrition, and public health. As noted above, however, mechanisms for including the social sciences in the ARS research agenda and stimulating appropriate collaboration pose a major challenge for the national agricultural research program. ARS and ERS will need to work together to identify ways in which this collaboration might successfully occur in the future.

In one of the unique collaborations in REE, particularly at the land-grant university level, knowledge has been generated through research and disseminated and applied through teaching and extension. No other scientific community enjoys a direct and formal relationship with a community-based educational organization committed to putting its knowledge and scientific findings to work to improve communities and citizens' lives. Collaboration has been strengthened by having extension faculty in university departments of agriculture, forestry, and human ecology and often by having faculty with joint research and extension appointments. Organizationally, at USDA, this collaboration was enhanced several years ago with the merger of the USDA Extension Service and the Cooperative State Research Service to form CSREES. The collaboration between

## BOX 5-1
## Research Partnerships in Which REE Has Provided Leadership

- A rainbow trout genomics project is coordinated by the ARS National Center for Cool and Cold Water Aquaculture (NCCCWA), Leetown, West Virginia. Participants in this national effort include scientists from the NCCCWA, West Virginia University, the University of Idaho, the University of Connecticut, Washington State University, and an ARS scientist at the University of Idaho's Hagerman Fish Culture Experiment Station. The purpose of this research partnership is to develop genetic tools and information that will lead to improved strains of farmed rainbow trout that display greater production efficiency and enhanced traits and product quality for consumers.

- Salt cedar is an exotic species that has invaded riparian communities of the western United States, disrupting native ecosystems and reducing the usefulness and value of these communities. To manage salt cedar, ARS is leading a national effort, via an IFAFS grant, that is exploring ecosystem response to the salt cedar beetle (a biologic control species). Members of the team include ARS scientists from Albany, California; Temple, Texas; and Davis, California; and other scientists from the University of California, University of Wyoming, and Texas A&M University. Study sites are in California, Wyoming, and Texas.

- ERS is part of a regional research project, Private Strategies, Public Policies, and Food System Performance (NE-165), that conducts collaborative research on the impacts of changes in strategies, technologies, consumer behavior, and policies on the economic performance of the food system and on how private and public strategies influence improvement in food safety and other quality attributes. It has over 100 members from around the world, primarily universities and government agencies. NE-165 developed reliable methods for measuring the private and public benefits and costs of improving the quality of food products. Examples include measuring the benefits of reducing foodborne illness and the costs of adopting new control methods such as hazard analysis and critical control points. NE-165 research has been used by companies in assessing food-quality programs and by Congress, USDA, and the Food and Drug Administration (FDA) in designing and evaluating policy. ERS economists have been strong participants and leaders in this collaborative effort.

research and extension has been highly successful in universities, but collaboration between extension and the other REE agencies—ARS, ERS, and the National Agricultural Statistics Service (NASS)—has not been as effective. There are no formal links between extension and the other three REE agencies.

## COLLABORATION IN THE FEDERAL GOVERNMENT

Increasingly, collaboration with other government agencies is important to REE's success in carrying out its mission. Many of the issues facing REE agencies require expertise and knowledge that extend beyond its traditional scope. Therefore, REE agencies have developed numerous collaborations with both federal research and action or regulatory agencies. The list of collaborators includes the National Science Foundation (NSF), the National Institutes of Health (NIH), EPA, National Aeronautics and Space Administration (NASA), FDA, National Oceanic and Atmospheric Administration (NOAA), the US Military, DOI, the Department of Energy (DOE), and the Department of Defense (DOD).

In nutrition, for example, existing NIH–ARS collaborations include the National Food and Nutrition Analyses Program, which sets priorities for maintaining the National Nutrient Databank; an interagency agreement between NIH and the ARS Food Composition Laboratory for the development of new chemical methods for analyzing nutrients and other biologically active compounds in foods; and a 1998 Carotenoid Food Composition Database developed jointly by ARS and the Nutrition Coordinating Center at the University of Minnesota with partial funding from the National Cancer Institute (NCI) (USDA, 1998). Other potential models of collaboration are the National Health and Nutrition Examination Survey, conducted by ARS and the Centers for Disease Control (CDC) National Center for Health Statistics (NCHS), and a partnership involving CDC, NIH, USDA, and others to improve availability of high-quality data related to fruit and vegetable consumption in support of the "5 A Day for Better Health Program."

There is still untapped potential for collaboration in food and nutrition. In the case of complex diseases with nutritional components, such as cardiovascular disease and osteoporosis, most genetic research is conducted under the auspices of NIH and private industry, but broader collaboration will be essential for addressing these health issues. Nutrition research at NIH includes the determination of the biochemical functions of nutrients and other food components in biologic systems, exploring differences in biochemical functions resulting from genetics, environmental factors, and disease conditions. NIH nutrition research focuses on how to prevent, control, and treat diet-related diseases and conditions. The results of NIH nutrition research could be considered by REE agencies in planning their research agendas, particularly those that involve selecting foods and food components for analysis for the National Nutrient Databank (ARS), applying results to community nutrition programs and determining nutrition-behavior interventions in community programs (CSREES), and statistical evaluation of the relevance of diet-related demographic variables (ERS and NASS). USDA had traditionally not been focused on diet-related disease but has more recently been conducting some research and community programs that concern obesity and diabetes. USDA's movement into these disease topics could be strengthened, and USDA could make more progress by having a thorough knowledge of NIH's past and present research in these topics and expanding into community programs (not traditionally done by NIH) or other topics not covered by NIH. Another potential collaborative research topic is the development of methods to assess the intake of dietary supplements by the American public. The NIH Office of Dietary Supplements (ODS) is communicating with ARS and NCHS in developing such methods.[1] This will require the development of a database on the composition of dietary supplements—a tremendous task, considering the huge number of products and different potencies.

In food-safety research, interagency collaboration has been critical and holds further possibilities; here collaborative research is the most cost-effective and timely mechanism for identifying critical control points and implementing intervention strategies. USDA's Foodborne Outbreak Response Coordinating Group—which links federal, state, and local government agencies to enhance coordination and communication in responding to outbreaks, uses resources efficiently, and prepares for new and emerging threats to the food supply—is a collaborative mechanism that could be useful in food-safety research.

Collaborative models exist in environmental research. The Sustainable Agriculture Research and Education program is an example of a collaborative model among USDA, EPA, and several profit and nonprofit organizations for research using a whole-systems perspective, a participatory approach, and a decentralized structure. In addition, for many years, ERS, NSF, EPA, and NOAA have coordinated their extramural funding (typically with universities) for climate-change research, especially addressing the socioeconomic impacts. This approach not only reduced duplication of research effort and improved efficiency of the funding process but led to improved planning and coordination of intramural and multidisciplinary research efforts. There is further room for development of collaborative efforts in environmental research. Many of the environmental research frontiers identified by this report overlap with issues facing other federal agencies, particularly the land and natural-resource management agencies in DOI. Examples include invasive species, environmentally sound management practices, carbon sequestration, and integration of spatial technologies and distributed datasets into decision-making for natural-resource management. At present, however, collaborative research between REE and DOI agencies is haphazard and usually involves small-scale, project-by-project funding of REE scientists by DOI

---

[1]ODS has previously worked with USDA's Food and Nutrition Service to develop the IBIDS database (*http://dietary-supplements.info.nih.gov/*) on research on dietary supplements.

agencies. Mechanisms that could enhance collaboration on shared environmental research problems include collaborative development of requests for proposal (RFP)s, shared planning of research initiatives and implementation of research findings on the ground, and congressional appropriations of research funds to DOI agencies. Such collaborative approaches could be used more widely with nongovernment organizations whose missions and goals overlap with those of USDA. Development of the National Management Plan under the recent executive order on invasive species required USDA, DOI, and other agencies to work collaboratively in building a research plan, one of several sections of the National Management Plan. The current administration has requested a cross-cut for 2004 budget requests from the agencies, which presumably could help to clarify where and how funds and resources will be used to meet shared needs. This approach provides one potential model for collaboration among agencies.

The national extension network was developed when the economy was primarily agricultural and the population predominantly rural. Today, the US economy is diverse, the population highly heterogeneous and urban, and the nation an integral part of a global economy and society. Therefore, the collaboration that extension pursues must be more flexible and adaptable and extend far beyond its traditional partners in REE (ECOP, 2002).

Other federal agencies, such as NIH and EPA, have expressed interest in using the extension system for their own outreach efforts, and there are numerous models of new partnerships between Cooperative Extension and federal agencies. For example, a CSREES-administered collaborative project, Healthy Indoor Air for America's Homes, links EPA with Cooperative Extension in 46 states to eliminate household hazardous substances. Extension could potentially facilitate the sharing of information and the coordination of research priorities between federal agencies (including DOD, DOE, NASA, NIH, and NSF) that address related research. That would reduce duplication of activities and maximize the use of resources. Such coordination could include priority-setting, cofunding of initiatives, and dedicated funding by single agencies. Strengthening the leadership of the current CSREES administrator and the REE undersecretary as advocates for extension and dissemination of the benefits of research could help to promote the outreach and educational opportunities in CSREES and the land-grant universities.

**FINDING: There is tremendous potential for collaboration and strategic alliances involving the Extension System inside and outside the university with ARS, NASS, ERS, and other federal agencies (such as DOC, DOD, DOE, EPA, and NIH) to address the social and economic issues facing all communities.**

One proposed new mechanism to strengthen collaboration among federal agencies has been the "virtual research and development centers," temporary task forces or teams in their home agencies or institutions. It has been suggested that REE develop this capacity and fund such approaches that would involve REE agencies as partners. The leader of an REE virtual R&D center would be empowered to recruit, organize, and coordinate the services of professionals in any of the four REE agencies, any other USDA agency, or elsewhere as needed, such as in government, university, or private-sector institutions. The members of the center could be based in their home institutions but would provide knowledge, advice, skills, and equipment as needed to accomplish the specific goals of their center. Possible challenges in implementation include complexity in using a matrix approach in a line organization, such as USDA. In addition, there is a strong tendency for centers, once established, to persist, often long past their useful life, so disbanding of the center when its mission is accomplished may pose a challenge. However, the committee believes that many national challenges will require sustained and creative efforts, which such virtual laboratories might best facilitate.

## INTERNATIONAL COLLABORATION

In the increasingly global economy and society, international collaboration of REE has expanded and will probably continue to do so. The public-sector agricultural research and extension community has a long tradition of international collaboration. Perhaps the most well-developed international scientific network has grown through the Consultative Group on International Agriculture Research (CGIAR). International agricultural-science collaboration was important for the latter part of the 20th century in responding to the rising economic growth and food needs of an expanding global population.

For most of the last century, international students and scholars in the agricultural, nutritional, environmental, and rural social sciences have generally constituted the largest international contingent attending US universities and collaborating in the federal laboratories. Table 5-1 shows the substantial and consistent investment by ARS in visiting international scientists. In FY 2001, visiting scientists made up almost 8% of the 1,980 scientists in the ARS workforce

**TABLE 5-1** Visiting Scientists at ARS, 1998–2001

|  | 2001 | 2000 | 1999 | 1998 |
|---|---|---|---|---|
| Number of visiting scientists | 156 | 129 | 135 | 193 |
| Number of countries represented | 15 | 36 | 48 | 41 |
| Top countries represented | China, Korea, Brazil | China, Italy, France | China, Japan, Brazil | China, Korea, Brazil |
| Cost to support each scientist | $303,247 | $122,940 | $150,000 | NA[a] |

[a]NA = data not available.
Source: USDA ARS (2002).

(USDA, 2001a). ERS also has hosted visiting scientists from transitioning and developing economies and details staff to international organizations, such as the Food and Agriculture Organization (FAO) and the Organization for Economic Cooperation and Development. NASS has an international data-collection unit that assists developing and transitioning economies in survey and census design and data-collection activities.

Many US colleges of agriculture, unlike other university colleges, have separate offices and associate deans for international programs that coordinate and promote international collaboration. The Office of International Programs at ARS is expanding its mission to increase memoranda of understanding with other countries regarding mutually beneficial agricultural research. It is likely that public-sector agricultural scientists have engaged in more international collaboration in more countries and locations than any other group of US scientists.

Historically, the emphasis on food and agriculture in the US Agency for International Development (USAID) has led to substantial funding for international agricultural collaboration. However, US investments in international agriculture have declined from $1.2 billion in 1985 to $332 million in 1999 (USAID, 1985, 2000), and funding for bilateral assistance in agricultural research in USAID declined from a peak of about $250 million in the 1980s to about $60 million in 1997 (in constant 1987 dollars; Alex, 1997). However, political efforts are under way to restore and expand USAID investments in agriculture and rural development. The need remains high for funding agricultural development and international agricultural collaboration. A major mechanism for USAID-university partnerships for international collaboration has been the Collaborative Research Support Program (for example, in sorghum and millet, bean and cowpea, livestock, aquaculture, and sustainable agriculture and natural-resource management research). These multidisciplinary, multiinstitutional, international research programs have been excellent models for international agricultural collaboration. They have focused on collaboration among developing national scientists and US public-sector agricultural scientists working on food, nutrition, and environmental issues critical to developing nations.

Other international collaborative efforts have involved CGIAR-international agriculture research centers (CGIAR-IARC; 16 worldwide), FAO, the World Bank, and numerous national agricultural research systems, such as Empresa Brasileira de Pesquisa Agropecuária in Brazil and the India Council for Agriculture Research.

Given the limitations on human and financial resources available in REE agencies, it may not be feasible or efficient to contribute directly to improving agricultural productivity in developing countries. It may be preferable to approach this contribution from another perspective. Public agricultural research has been and continues to be important in agricultural-productivity growth and enhancing food security in developing countries. Alston (2002) reviews and summarizes several studies that conclude that about half the research benefits in any nation may be due to spillover effects from research conducted in other countries and that spillover effects may benefit other countries as much as the nation conducting the research. Every study reviewed found important international spillover effects of agricultural R&D. Furthermore, varietal-improvement R&D from the CGIAR centers demonstrates large benefits at the country level as well as globally. For example, Pardey et al. (1996) showed that US contributions to research on rice and wheat improvements in the CGIAR have had substantial returns to US agriculture.

Although agricultural spillovers are important to individual countries and in the aggregate, the process through which these spillovers affect individual and global productivity improvements is not well understood. On the basis of varietal-improvement case studies and the more aggregate spillover measures, USDA scientists may already be making a substantial indirect contribution through spillover benefits to developing-country productivity and food security. Alston (2002) indicates that this spillover contribution could be enhanced by additional bilateral arrangements with individual countries and multilateral arrangements with CGIAR and FAO.

There is a growing gap in agricultural research intensity (agricultural R&D relative to national agricultural gross domestic product) between developed and developing countries (see Chapter 4). In 1995, developing countries expended only $0.62 on public agricultural R&D per $100 of agricultural GDP, whereas developed countries expended $2.64 (Pardey and Beintema, 2001). This further suggests a great potential benefit for developing countries resulting from collaboration with developed countries' institutions that would take advantage of under-exploited opportunities in R&D. Collaborating institutions could be regarded as a mechanism for providing international public R&D research goods and services as opposed to a mechanism for transferring humanitarian development aid. New possibilities are emerging as a consequence of concerns about national and international security, bioterrorism, and international education. USDA could play a role in fostering such collaborations.

ARS has a long history of cooperation with the CGIAR-IARC system with more than 25 formal and informal research collaborations that cover a wide array of topics in material-resources management, access to and exchange of global germplasm, disease and pest management, and enhancing crop and animal productivity and quality. However, most of those efforts remain underfunded. Indeed, CGIAR donor support has dramatically decreased in the last 2 years. Collaboration with the National Agricultural Research Institutes could also be strengthened. ARS and US universities have inadequate funds to devote to international activities. At the same time, the high quality of science around the globe and the increasing interconnectedness of the food and fiber system worldwide requires more, rather than less, international collaboration. It will take aggressive and creative efforts to strengthen existing collaboration and to identify and pursue new collaboration among the universities, governments, and the private sector.

To conclude, there is evidence that REE has a strong history of working collaboratively. Strengthened collaboration with new and existing partners holds promise for addressing the complexity of issues and challenges facing the global agricultural system and for engaging in the new research opportunities described in this report.

## COLLABORATION WITH THE PRIVATE SECTOR

During the last 20 years, the convergence of a number of political, economic, social, scientific, and technologic developments has affected how agricultural science is conducted and commercialized and the evolution of new institutional collaboration and public and private research partnerships. The new commercial opportunities; patent laws and decisions (such as the 1980 US Supreme Court decision in *Diamond v. Chakrabarty*, 1980, extended by the 2001 US Supreme Court decision that seeds and seed-grown plants can be patented in *J.E.M. Ag Supply, Inc. v. Pioneer Hi-Bred International, Inc.*); federal policies (such as the Government Patent Policy [Bayh-Dole] Act of 1980 [US Congress, 1980] and the Federal Technology Transfer Act of 1986 [US Congress, 1986]); establishment of minimal standards of intellectual-property protection,[2] mechanisms for intellectual-property rights enforcement, and provisions for dispute settlement for World Trade Organization (WTO) members under the TRIPS (trade-related aspects of intellectual property) Agreement (WTO, 1994); growth in private-sector research; and a relative decline in public-sector funding of agricultural research have all contributed to a changing collaborative relationship between universities and industries (Josling, 2001; Murashige, 1997; Parker et al., 2001). The new types of university–industry collaboration are generally more varied, of wider scope, more aggressive and experimental, and more publicly visible than past relationships. They involve diverse approaches that include large grants and contracts between companies, universities, and government laboratories in exchange for patent rights to and exclusive licenses of discoveries; programs and centers organized with industrial funds at major universities (now totaling over 1,000), which give participating private firms privileged access to resources and a role in shaping research agendas; professors, particularly in the biomedical sciences, serving in extensive consulting capacities on scientific advisory boards or in managerial positions in firms; faculty and research scientists receiving research funds from private corporations in which they hold substantial equity; and public universities and government laboratories establishing business startups and for-profit corporations to develop and market innovations arising from research.

The Technology Transfer Act of 1986 (US Congress, 1986) established the cooperative research and development agreement (CRADA), a mechanism

---

[2]Intellectual-property protection includes patents, copyrights, plant-variety protection certificates (Plant Variety Protection Act [US Congress, 1970]), trademarks, copyrights, and technology licenses.

through which federal and nonfederal researchers could collaborate (Adams et al., 2001; Fuglie et al., 1996; Huffman and Just, 1999a). The principal objective of a CRADA is to link the research capacity of federal laboratories with the commercial research and marketing expertise of the private sector. Under a CRADA, a federal laboratory may provide personnel, equipment, and laboratory privileges for commercial activity. Similarly, the private-sector collaborator may contribute funds directly to the federal laboratory in return for the right of first refusal to negotiate an exclusive license of any joint discovery and may be given exclusive access to data from a joint project. In addition to CRADAs, there are other arrangements for private-sector collaboration, such as trust-fund agreements, research instruments in which a private-sector cooperator is not offered a first right of refusal to negotiate an exclusive license; patent licensing, in which public entities patent inventions and then grant exclusive, limited exclusive, or non-exclusive licenses to private companies to use or market the inventions; and research consortia, in which several institutions undertake joint research with or without a private-sector partner (USDA, 2000).

CRADA activity increased rapidly after 1987 (see Table 5-2), and in 2000 over 250 CRADAs were active, using combined public and private resources of

**TABLE 5-2** USDA Technology-Transfer Activities, 1987–2000

| Year | Number of Patents Awarded | Patent License Royalties, millions of dollars | Number of Active CRADAs with Private Sector | Value of CRADAs,[a] millions of dollars |
|---|---|---|---|---|
| 1987 | 34 | 0.09 | 9 | 1.6 |
| 1988 | 28 | 0.10 | 48 | 8.7 |
| 1989 | 47 | 0.42 | 86 | 15.6 |
| 1990 | 42 | 0.57 | 145 | 18.9 |
| 1991 | 57 | 0.83 | 181 | 25.6 |
| 1992 | 56 | 1.00 | 172 | 30.0 |
| 1993 | 57 | 1.50 | 172 | 34.0 |
| 1994 | 40 | 1.40 | 208 | 61.3 |
| 1995 | 38 | 1.60 | 229 | 80.1 |
| 1996 | 53 | 2.10 | 244 | 98.9 |
| 1997 | 35 | 2.30 | 273 | 155.5 |
| 1998 | 57 | 2.40 | 271 | 120.2 |
| 1999 | 74 | 2.40 | 298 | 136.7 |
| 2000 | 64 | 2.60 | 257 | 125.1 |

[a]Includes total value of USDA and private-sector resources committed to active CRADAs over their lifetime.

Source: USDA, ERS, compiled from ARS Office of Technology Transfer data in USDA (US Department of Agriculture). 2000. Agricultural Resources and Environmental Indicators, 2000. Washington, D.C.: Economic Research Service, Resource Economics Division, US Department of Agriculture. Available online at *http://www.ers.usda.gov/Emphases/Harmony/issues/arei2000/*.

$125 million. ARS contributions are on the average about one-third of total resources and thus less than 5% of the ARS budget (Day-Rubenstein and Fuglie, 1999). In FY 2001, ARS inhouse contributions were about $4.2 million, representing 35% of the total contributions. Cooperator inhouse contributions ($6 million, or 50% of the total contributions) and cooperator contributions paid to ARS ($1.9 million, or 15% of the total) accounted for the remainder. A number of patents have been awarded, and the patenting and licensing royalties returning to ARS are now $2.6 million per year. According to ARS's Office of Technology Transfer, in 2001, of the total royalties, 26% is allocated to incentive awards to inventors (the law requires a minimum of 15%), 41% supports the salaries of some of the technology-transfer staff to facilitate more agreements, and 27% supports patent filing preparation, fees, and patent annuity payments (USDA, 2001d).

The CRADA seems to be an important policy tool for increasing technology transfer. Adams et al. (2001) examined industrial research and found that CRADAs dominate the channels of technology transfer from federal laboratories to the private sector, largely because of the effort that they demand of both parties. Since the CRADA legislation was enacted in 1986, there has been increased spending by the private sector in federal laboratories. Public–private partnerships are less well developed in agricultural research than in industrial research and account for a smaller share of total research resources. But the existence of CRADAs to help shape public–private collaboration since 1987 has resulted in many successful examples of technology transfer. Box 5-2 provides examples of CRADA activities that have resulted in important innovations in agricultural production, environmental protection, and human health.

An important question in collaboration with the private sector is how the results of research are controlled and shared. ARS was delegated authority by the secretary of agriculture to administer the patent and license programs for USDA. In contrast, CSREES subordinates its intellectual-property governance to that of the institution receiving funds. The ARS Office of Technology Transfer is assigned the responsibility for protecting intellectual property, developing strategic partnerships with outside institutions, and performing other appropriate functions that enhance the effective transfer of ARS technologies to users. The Office of Technology Transfer also ensures that information about the commercial successes of ARS is made available to the public. The stated ARS policy is "to use the patent system to promote the utilization of inventions arising from its research, to ensure that sufficient rights in inventions are obtained to meet the needs of the Government, and to bring the invention to practical application." That is an extremely broad policy statement that should be rewritten to give specific guidance on USDA patent policy to other constituents in the agricultural research community (such as industry).

The number of patents and the licensing and royalty fees generated provide some measure of the effectiveness of technology-transfer efforts at USDA (Table 5-2). This kind of research is relatively new, so no system is in place for performance measurement in terms of technology transfer, adoption, and impact. Future monitoring of the impact of private-sector collaboration would help to assess the benefits of such arrangements, which would help to address concerns about such collaboration.

## Benefits of and Concerns about Public–Private Collaboration

The outcomes of collaboration between the two distinct and complementary research communities can be both favorable and adverse. First, ARS, land-grant university, and industry collaboration may bring useful products to market more rapidly and promote US technologic leadership in a changing world economy (Reilly and Schimmelpfennig, 2000). Second, in light of funding stagnation in USDA and in many cases at the state level, such collaboration is a means of raising new funds for public research, graduate education, and postdoctoral fellowships (Smith et al., 1999). Third, the collaboration can introduce public-sector scientists and students to industry and enhance their understanding of the nonacademic world of science (Rogers and Bozeman, 2001). Fourth, the joint efforts may expand the scientific network, increasing communication among some industry, ARS, ERS, and university scientists to provide them access to cutting-edge research tools, proprietary materials, and vast databases owned by particular companies (Shoemaker et al., 2001).

A number of concerns have been voiced regarding these new relationships. First and foremost, there is concern that private-sector funds will set public-sector priorities and divert public resources from research topics with broad social benefits (Feller et al., 2002; Parker et al., 2001). If a sufficiently large and influential number of academic scientists and engineers become involved with industry, a whole range of research agendas that are traditionally the purview of the public sector might be de-emphasized (Huffman and Just, 1999b; Lacy, 2001). The scientific community might become desensitized to the environmental or social impacts of proprietary research. Second, long-term research, previously a major emphasis of the public sector, may decline. Dependence on private-sector funds will generally change not only the time frame but also the stability of funding (Shoemaker et al., 2001). It seems unlikely that the public-sector-industry relationships will provide stable long-term funding, nor will they substantially address the capital needs of the public sector. Third, there are concerns about restricting scientific communication or the possibility of shelving research of interest to the public but not to corporations (Heller and Eisenberg, 1998; Lacy, 2001). Fourth, there is concern about how the funds generated by royalty income may be allocated to current research and reserves for future research (Dasgupta and David, 2002). Finally, a dominant problem that public agencies face is gaining access to proprietary technologies, an issue particularly relevant to the ability to execute and commercialize research that is at least partially predicated on other technologies that are legally sequestered by other organizations—so-called "interlocking"

## BOX 5-2
## Collaborative Activities Through Cooperative Research and Development Agreements (CRADAs)

The following are examples of important agricultural and biologic research problems that have been or are being addressed through CRADAs.

**Poultry Vaccination**

In 1987, ARS licensed to Embrex, Inc. a new technique for immunizing poultry by injecting vaccines into their eggs. Thereafter, a CRADA was established between ARS and Embrex, which at the time was a small company consisting of two staff members. After entering into this agreement, Embrex proceeded to design and obtain a patent for a machine capable of vaccinating a substantial number of hatchery eggs per hour. This method of inoculation now safeguards the majority of broiler chickens in North America, and Embrex has grown into a corporation that employs over 120, serving not only domestic markets but also international markets through its overseas operations. Embrex has patented three inventions that originated in ARS (USDA, 2001c).

**Low Phytic Acid Grains**

As livestock feed, grain that is low in phytic acid offers possible benefits for nonruminant nutrition in addition to the potential environmental benefit of lowering the amount of phosphate in runoff from livestock farms. In 1994, ARS submitted a patent for a technique used to produce strains of corn and other crops that would be low in phytic acid. By 1997, ARS was issued a patent for this method; but before it could be commercialized, further research was needed to incorporate the expression of low phytic acid into hybrid corn. Recognizing the need for research, ARS and Pioneer Hi-Bred (a developer and supplier of advanced plant genetics) established a CRADA. Thus far, three nonexclusive licenses for the technology have been granted, and research has indicated a considerable amount of progress in developing hybrid corn low in phytic acid. In addition to hybrid corn, low-phytate barley, rice, and soybeans have attracted attention for research with the potential for future commercialization (USDA, 2001c).

**Protection of Cacao Trees**

Millions of Latin American, African, and Asian farmers have suffered serious economic losses due to three fungal diseases: black pod rot, frosty pod rot, and witches' broom. Almost 3 years ago, ARS participated in an international collaboration of research groups—including the American Cocoa Research Institute, M&M Mars, Inc., and the Brazilian Cacao Authority—to explore beneficial fungi that would inhibit fungi harmful to cacao beans. Previous ARS research resulted in the commercialization of a product based on beneficial fungi used to control diseases in some fruits and vegetables. ARS's Biocontrol of Plant Diseases Laboratory in Beltsville, Maryland, later entered into the collaborative research effort with M&M Mars and other international partners (USDA, 1999).

In 2001, ARS's Subtropical Horticulture Research Station in Miami, Florida, and M&M Mars entered into a CRADA concentrating on the development of cacao trees with greater resistance to fungal diseases. ARS has discovered several cacao genes that confer disease resistance (USDA, 2001b).

**Taxol**

Taxol is an anticancer drug obtained from the bark of the Pacific yew, a tree that, although rare, is an important natural resource. Through an agreement between USDA and NCI several years ago, results of research on the properties of taxol led both agencies to conclude that its benefits warranted commercialization. After instituting a competitive-award system, NCI and the pharmaceutical company Bristol-Myers Squibb entered into a CRADA, providing the company with access to NCI clinical data on taxol. In 1992, taxol was approved by FDA as an ovarian-cancer drug. Although Bristol-Myers Squibb, USDA, and DOI established an agreement whereby the company was given the sole right to use Pacific yew trees growing on federal land, Bristol-Myers Squibb was made responsible for studying alternatives to using yew trees for obtaining taxol. Collective research has led to various alternative means for producing taxol, and Bristol-Myers Squibb has ceased relying on yew trees that grow on federal land.

As a result of Bristol-Myers Squibb's agreement with NCI, further agreements between the company and other institutions—both public and private—were established. Such agreements enabled the collection and synthesis of a considerable amount of research data. Moreover, because NCI encouraged additional and alternative examinations of taxol or drugs similar to taxol, it is evident that both public and private institutions were able to participate in this endeavor without being excluded from conducting relevant research and without other firms' being excluded from the anticancer-drug market (USDA, 1996).

technologies. This is an unfortunate characteristic of today's leading-edge agricultural research, especially in biotechnology (Nottenburg et al., 2002).

Because public–private collaboration is relatively new to the agricultural sector and these relationships are still evolving, there are many unanswered questions about their benefits and risks. As we note in Chapter 3, the management of intellectual property in agriculture constitutes an important opportunity for future research. The data-gathering and analysis being carried out by ERS for monitoring public and private research and development represent a very important resource. Such socioeconomic research can help to inform future policy at ARS and help REE to provide leadership for land-grant universities as they develop technology-transfer models further. Examination of existing models in other fields with long-established public–private collaboration, such as colleges of engineering, can also help the public agricultural sector to define policy (Feller et al., 2002; Rogers and Bozeman, 2001).

**FINDING: Collaboration between the public and private sectors is increasing in agricultural research. Benefits of such collaborations include more-successful technology transfer, increased support for research, and expanded scientific networks. Concerns about such collaborations include their potential effect on priority-setting in the public sector, on scientific-information generation, and on the allocation of resources for future research. Many questions regarding the management of intellectual property in agriculture are unresolved, and policy is not well defined.**

## FUTURE STRATEGIES TO MANAGE PUBLIC–PRIVATE COLLABORATION

The future will depend on strong, independent, complementary research efforts by the public sector and the private sector. Neither will thrive for long if the other is weakened or its goals and integrity are eroded. The future will also involve continued expansion of public-sector and industry relationships and new and creative forms of collaboration. REE can play a leadership role in the public agricultural-research system in helping to define relationships that will best serve the public interest.

**RECOMMENDATION 6: REE should provide national leadership in developing intellectual-property policy for agricultural research. REE should address the potential consequences of public–private collaboration with appropriate policies, practices, and organizational arrangements that**

- Promote the greatest public benefit from agricultural research.
- Protect the public investment in research.
- Prevent diversion of public resources away from research that can be carried out only in the public sector.
- Pursue strategic private-sector collaboration necessary to achieve public goals.

To accomplish these objectives, REE should establish ways to measure the effectiveness of technology generation and transfer through private-sector collaboration.

The policy should broadly define the extent to which collaboration would involve support from the private sector and how earnings from successful technologies could be reinvested in research programs. The committee acknowledges that, in practice, implementation of intellectual-property policy is a complex and often case-specific undertaking, as are the implications of intellectual-property policy for research.

## SUMMARY

This chapter has considered collaboration and strategic alliances as avenues with great potential for addressing the research frontiers laid out in Chapter 3. Multidisciplinary, systems-level approaches were discussed as a complement to disciplinary approaches in addressing increasingly complex research problems. Examples of effective multidisciplinary and collaborative efforts of the REE agencies were described. The evolving relationship between the public and private sectors resulting from changes in policy and in science and technology was outlined, as were benefits of and concerns about public–private collaboration. A more comprehensive strategy to manage collaboration with the private sector is needed, including policies, practices, and organizational arrangements that consider the potential consequences of public–private collaboration for the public good.

## REFERENCES

Adams, J.D., E.P. Chiang, and J.L. Jensen. 2001. The Influence of Federal Laboratory R&D on Industrial Research. September. Gainesville, FL: The University of Florida, Department of Economics.

Alex, G. 1997. USAID and Agricultural Research: Review of USAID Support for Agricultural Research 1952–1996. Environmentally Sustainable Development Agricultural Research and Extension Group–Special Report No. 3. Washington, DC: World Bank.

Alston, J. 2002. Spillovers. Australian Journal of Agricultural and Resource Economics 46(3):315–346.

Amerman, R.C., G. Larson, and M. O'Neill. 2001. USDA Water Quality Initiative. Presentation to National Research Council Committee on Opportunities in Agriculture Public Workshop, May 22–23. Washington, DC.

Busch, L., and W.B. Lacy. 1983. Science, Agriculture, and the Policies of Research. Boulder, CO: Westview Press.

Caswell, J. 2001. Economic approaches to measuring the significance of food safety in international trade. International Journal of Food Microbiology 62(3):261–266, Special Issue, Dec. 20, 2000.

Dasgupta, P., and P.A. David. 2002. Toward a New Economics of Science. Pp. 219–248 in Science: Bought and Sold, P. Mirowski and E. Sent, eds. Chicago: University of Chicago Press.

Day-Rubenstein, K., and K.O. Fuglie. 1999. Resource allocation in joint public–private agricultural research. Journal of Agribusiness 17(2):123–134.

*Diamond v. Chakrabarty.* 1980. 447 US 303.

ECOP (Extension Committee on Organization and Policy). 2002. The Extension System: A Vision for the 21st Century. Washington, DC: National Association of State Universities and Land-Grant Colleges.

Feller, I., C.P. Ailes, and J.D. Roessner. 2002. Impacts of research universities on technological innovation in industry: Evidence from engineering research centers. Research Policy 31:457–474.

Flora, C.B., ed. 2001. Interactions Between Agroecosystems and Rural Communities. Boca Raton, FL: CRC Press.

Fuglie, K., N. Ballenger, K. Day, C. Klotz, M. Ollinger, J. Reilly, U. Vasavada, and J. Yee. 1996. Agricultural Research and Development: Public and Private Investments under Alternative Markets and Institutions. USDA-ERS, Agricultural Economics Report No. 735, May. Washington, DC: Economic Research Service, US Department of Agriculture.

Heller, M.A., and R.S. Eisenberg. 1998. Can patents deter innovation? The anticommons in biomedical research. Science 280:698–701.

Huffman, W.E., and R.E. Evenson. 1993. Science for Agriculture: A Long Term Perspective. Ames, IA: Iowa State University Press.

Huffman, W.E., and R.E. Just. 1999a. Agricultural research: Benefits and beneficiaries of alternative funding mechanisms. Review of Agricultural Economics 21(Spring /Summer):2–18.

Huffman, W.E., and R.E. Just. 1999b. The organization of agricultural research in western developed countries. Agricultural Economics 21:1–18.

*J.E.M. Ag Supply, Inc. v. Pioneer Hi-Bred International, Inc.* 2001. 534 US 124.

Josling, T. 2001. International institutions, world trade rules, and GMOs. Pp. 117–130 in Genetically Modified Organisms in Agriculture: Economics and Politics, G.C. Nelson, ed. San Diego, CA: Academic Press.

Lacy, W.B. 2001. Generation and commercialization of knowledge: Trends, implications and models for public and private agricultural research and education. Pp. 27–54 in Knowledge Generation and Technical Change: Institutional Innovation in Agriculture, S.A. Wolf and D. Zilberman, eds. Boston: Kluwer Academic Publishers.

Matson, P.A., W.J. Parton, A.G. Power, and M.J. Swift. 1997. Agricultural intensification and ecosystem properties. Science 277:504–509.

Murashige, K.H. 1997. Patents and biotechnology. Pp. 283–290 in AAAS Science and Technology Policy Yearbook. A.H. Teich, S.D. Nelson, and C. McEnaney, eds. Washington, DC: American Association for the Advancement of Science.

Nottenburg, C., P.G. Pardey, and B.D. Wright. 2002. Accessing other people's technology for nonprofit research. Australian Journal of Agricultural and Resource Economics 48(3):389–416.

Pardey, P.G., and N.M. Beintema. 2001. Slow Magic: Agricultural R&D a Century after Mendel. Washington, DC: International Food Policy Research Institute.

Pardey, P.G., J.M. Alston, J.E. Christian, and S. Fan. 1996. Hidden Harvest: US Benefits from International Research Aid. Washington, DC: International Food Policy Research Institute.

Parker, D., F. Castillo, and D. Zilberman. 2001. Public–private sector linkages in research and development: The case of US agriculture. American Journal of Agricultural Economics 83(3):736–741.

Reilly, J.M., and D.E. Schimmelpfennig. 2000. Public–private collaboration in agricultural research: The future. In Public–Private Collaboration in Agricultural Research: New Institutional Arrangements and Economic Implications, K.O. Fuglie and D.E. Schimmelpfennig, eds. Ames, IA: Iowa State University Press.

Rogers, J.D., and B. Bozeman. 2001. Knowledge value alliances: An alternative to the R&D project focus in evaluation. Science, Technology, and Human Values 26(1):23–55.

Shoemaker, R., J. Harwood, K. Day-Rubenstein, T. Dunahay, P. Heisey, L. Hoffman, C. Klotz-Ingram, W. Lin, L. Mitchell, W. McBride, and J. Fernandez-Cornejo. 2001. Economic Issues in Agricultural Biotechnology, ERS Agriculture Information Bulletin No. 762. March. Washington, DC: Economic Research Service, US Department of Agriculture.

Smith, K.R., N. Ballenger, K. Day-Rubenstein, P. Heisey, and C. Klotz-Ingram. 1999. Biotechnology research: Weighing the options for a new public–private balance. Pp. 22–25 in Agricultural Outlook, October 1999. Washington, DC: Economic Research Service, US Department of Agriculture.

USAID (US Agency for International Development). 1985. Title XII Report to Congress, FY 1985. Washington, DC: US Agency for International Development.

USAID (US Agency for International Development). 2000. Title XII Report to Congress. Washington, DC: US Agency for International Development.

US Congress. 1946. 7 USC. 1621–1627. Agricultural Marketing Act of 1946.

US Congress. 1970. P.L. (Public Law) 91-577. Plant Variety Protection Act of 1970.

US Congress. 1980. P.L. (Public Law) 96-517. Government Patent Policy Act of 1980.

US Congress. 1986. P.L. (Public Law) 99-502. Federal Technology Transfer Act of 1986.

US Congress. 1998. P.L. (Public Law) 105-185. Agricultural Research, Extension, and Education Reform Act of 1998.

USDA (US Department of Agriculture). 1996. Agriculture Research and Development: Public and Private Investments Under Alternative Markets and Institutions. Agricultural Economics Report No. 735. May. Washington, DC: Economic Research Service, US Department of Agriculture. Available online at *http://www.ers.usda.gov/publications/aer735/*.

USDA (US Department of Agriculture) 1998. USDA–Nutrition Coordinating Center Carotenoid Database for US Foods. Washington, DC: Agricultural Research Service, US Department of Agriculture. Available online at *http://www.nal.usda.gov/fnic/foodcomp/Data/car98/car98.html*.

USDA (US Department of Agriculture). 1999. Potential Chocolate Shortage May Be Foiled by Beneficial Fungi. ARS News and Information. October 25. Washington, DC: Agricultural Research Service, US Department of Agriculture. Available online at *http://www.ars.usda.gov/is/pr/1999/991025.htm*.

USDA (US Department of Agriculture). 2000. Agricultural Resources and Environmental Indicators, 2000. Washington, DC: Economic Research Service, Resource Economics Division, US Department of Agriculture. Available online at *http://www.ers.usda.gov/Emphases/Harmony/issues/arei2000/*.

USDA (US Department of Agriculture). 2001a. ARS Scientist Workforce. June 10. Agricultural Research Service, Office of Human Resources. Washington, DC: Agricultural Research Service, US Department of Agriculture.

USDA (US Department of Agriculture). 2001b. Resistance Genes Key to Protecting Chocolate Supply. ARS News and Information. October 15. Washington, DC: Agricultural Research Service, US Department of Agriculture. Available online at *http://www.ars.usda.gov/is/pr/2001/011015.htm*.

USDA (US Department of Agriculture). 2001c. Technology Transfer Through Cooperative Research and Development Agreements. Office of Technology Transfer, Agricultural Research Service. Washington, DC: Agricultural Research Service, US Department of Agriculture.

USDA (US Department of Agriculture). 2001d. USDA FY 2001 Annual Technology Transfer Report. Office of Technology Transfer, Agricultural Research Service. Washington, DC: Agricultural Research Service, US Department of Agriculture.

WTO (World Trade Organization). 1994. Agreement on Trade-Related Aspects of Intellectual Property Rights (TRIPS). Annex 1C of the Marrakesh Agreement Establishing the World Trade Organization. Marrakesh, Morocco. April 15. Available online at *http://www.wto.org/wto/english/tratop_e/trips_e/t_agm0_e.htm*.

Zucker, L.A., and L.C. Brown, eds. 1998. Agricultural Drainage: Water Quality Impacts and Subsurface Drainage Studies in the Midwest. Ohio State University Extension Bulletin 871. Wooster, OH: The Ohio State University.

# 6

# Quality and Impact Assurance in the REE Agencies

Ensuring the high quality, or excellence, of research is a key element of research management. But in mission-oriented research, such as that conducted in the US Department of Agriculture (USDA) Research, Education, and Economics (REE) mission area, impact assurance is equally important. Both quality assurance and impact assurance are necessary follow-up activities for relevance assurance, which was discussed in Chapter 4. The REE agencies have implemented several changes to strengthen quality assurance during the last decade. Measuring impacts is more difficult but has been accomplished for some dimensions. This chapter considers quality and impact assurance processes in the REE agencies, evidence regarding research impact, and how the processes can be improved to address impacts of the research opportunities outlined in Chapter 3.

## QUALITY ASSURANCE

Research quality[1]—the degree of excellence of research compared with other work being conducted in a field—rests on a foundation of scientific merit. Research quality is best evaluated by professional peers selected for their exper-

---

[1] According to the National Academies' Committee on Science, Engineering, and Public Policy, "there are at least two aspects of quality—one absolute and one relative. The absolute aspects are related to the quality of the research plan, the methods by which it is being pursued, its role in education when conducted at a university, and the importance of its results to its sponsor, either obtained or expected. The relative aspects pertain to its leadership at the edge of an advancing field. Although the leadership aspect is generally important, the results might in some cases be of great importance to an agency albeit not at the leading edge of a field" (NRC, 1999a).

tise in the field being assessed, experience, and objective judgment. This approach is known as peer review. Peer-reviewed science includes the peer review of discoveries before publication or patenting and the evaluation of quality of a research proposal, including the creativity of the idea, the technical soundness and appropriateness of the experimental design, the relationship to other scientific results from the literature, the record of the scientist or the scientific team, and the likelihood of scientific advances or practical applications or impacts. Peer review of research proposals is commonly used in allocating competitive funding to ensure quality in research design. Quality assessment of research outputs can include peer review of manuscripts, discoveries, research programs, and the quality of the publication in which the research results are published. Research quality depends heavily on research inputs, including the quality of the scientists conducting the research.[2] Rigorous evaluation systems for the scientists conducting the research and incentive structures for rewarding high-quality work and creative, independent thinking can contribute to ensuring high-quality outcomes.

### REE Quality-Assurance Mechanisms

The committee considered the REE agencies in light of metrics of quality and mechanisms for quality assurance. Generally, the committee identified a variety of quality-review and evaluation processes that are in place for all research projects and programs in the REE system; they are summarized in Table 6-1.

According to the strategic plan and performance plans of ARS (USDA, 1999, 2000a), each of roughly 1,100 research projects undergoes external merit-based peer review before new or renewed activities are begun. To ensure quality in the workforce, all ARS employees, including the scientific employees, are subject to annual performance reviews, and permanent scientists undergo a review of their progress on a 3- to 5-year cycle (discussed later in this chapter). A series of national program[3] reviews is designed to ensure the quality, relevance, effectiveness, and productivity of the work being done in each national program.

As mandated by the 1998 Agricultural Research, Extension, and Education Reform Act (AREERA) (US Congress, 1998), CSREES works with state partners receiving formula funds to develop 5-year plans of work, which are reviewed

---

[2]The committee notes that some private-sector institutions organize research to yield high-quality outcomes by emphasizing the hiring of high-quality researchers. For example, at 3M, high-quality staff are hired and given an ample endowment and flexibility to work on anything that is of interest to them (Arndt, 2002).

[3]ARS has recently restructured how it organizes and manages its national research programs (USDA, 1999). It has aggregated its research projects into 22 national programs that are guided by multidisciplinary teams of national program leaders. The national programs focus the work of the agency on reaching the goals defined in the ARS strategic plan.

**TABLE 6-1** Summary of REE Quality-Assurance Mechanisms

| Agency | Mechanism |
|---|---|
| Agricultural Research Service (ARS) | • Office of Scientific Quality Review (established in 1999)<br>• External peer review of projects before implementation<br>• Five-year-cycle review of national programs<br>• Annual reviews (project level and national-program level)<br>• Location reviews (research units)<br>• Review of quality of individual scientists—annual performance reviews for all ARS employees and 3- to 5-year Research Position Evaluation System peer reviews for senior scientists<br>• Solicitation of input on quality of ARS science from peer scientists and users |
| Cooperative State Research, Education, and Extension Service (CSREES) | • Annual review of individual projects<br>• Annual review of research programs<br>• Peer review of research proposals for competitive programs<br>• Review of output from and input into special grants (earmarks)<br>• External program review of the National Research Initiative by the National Research Council (NRC, 2000) |
| Economic Research Service (ERS) | • Peer review (internal and external) of all published material<br>• Rewards for productivity and high-quality performance (cash awards and promotions)<br>• Internal peer-review system for social-science positions, the Economist Position Classification System (established in 2000)<br>• External program review by the National Research Council (NRC, 1999b) |
| National Agricultural Statistics Service (NASS) | • Accuracy review using historical track records that compare forecasts with final, market-derived numbers for production-related reports; analysis of sampling errors and nonsampling errors<br>• Analysis of each step of data collection, processing, and estimation of statistics to evaluate the quality and accuracy of NASS reports<br>• Comparison of estimates with data sources outside the agency<br>• External technical review |

Source: Data provided by REE agencies, 2001.

annually for quality (USDA, 1998). In addition to quality review, these annual reviews also address accomplishments of research with respect to strategic goals and objectives, multistate activities, integrated research and extension, joint activities, and stakeholder input. CSREES is working to jointly establish and implement formal program-evaluation protocols, including expert assessments, with university and other partners and collaborators, through its Office of Planning and Accountability (USDA, 2000b). Reviews of academic departments at land-grant universities are also required by CSREES and occur on a 5-year cycle.

CSREES-administered competitive-grants programs, such as NRI and IFAFS, subject proposals for individual awards to an external merit-review process. CSREES post-award management procedures ensure that funds are expended according to the proposed plan of work, that progress reports are received and published in the Current Research Information System (CRIS) database, and that site visits and other oversight measures are performed as necessary. CSREES also increasingly requires its awardees to present results of their work at national and international scientific symposia. In addition to individual award evaluation, the programs are reviewed for quality as a whole each time a new request for abstracts is published. Special grants awarded to a particular institution can be reviewed before funds are awarded, and outputs of the special grant research can be subjected to peer review. An external review of a CSREES-administered competitive-grants program, the National Research Initiative, was conducted by the National Research Council in 2000 (NRC, 2000).

According to its annual performance plan, strategic plan, and annual performance report (USDA, 2000e, 2000f, 2001b), ERS systematically evaluates the quality of its work and considers the factors that affect quality. As part of that process, ERS conducts peer reviews before analysis is released, and the agency's successful contributions to professional conferences and journals test the appropriateness and rigor of the research methods in its analyses with respect to disciplinary standards. The National Research Council provided oversight for a 2-year review of the ERS program that was completed in 1998 (NRC, 1999b). In response to the report recommendations, ERS has taken a number of important actions, including the creation of an internal peer-review system. On the basis of recommendations of the Research Council study, ERS conducts broad reviews of critical aspects of its programs. It also initiated a collaborative university–ERS effort to measure the impacts of social-science research. The results of this analysis prove helpful to the agency in considering how to measure impacts and thereby the quality of its research.

According to its strategic plan and performance reports, NASS relies heavily on customer satisfaction surveys and end-user meetings—more related to relevance assurance than to quality assurance—to assess products and services. (See Chapter 4 for a discussion of relevance assurance mechanisms.) NASS reports that it uses historical track records to compare crop estimates and forecasts published during the growing season with the end-of-season final estimates to evaluate the accuracy of crop estimates and forecasts. NASS also uses external data sources to check the accuracy of its estimates. The NASS strategic plan reports that internal analysis of each step of data collection, processing, and estimation of production and price statistics is conducted to ensure quality (USDA, 2000g) and that professional standards are used in all major survey activities, including sampling frame development, sample design, questionnaire design and pretesting, data collection, analysis of sampling and coverage errors, nonresponse analysis, imputation of missing data, weighting, and variance estimation (USDA,

2002a). NASS does hire consultants from academe as consultants on data-collection issues, and periodic reviews by outside panels of technical experts are used to evaluate quality and reproducibility (USDA, 2001d, 2002a). However, it is not clear from the strategic planning and performance documents that these external reviews are conducted regularly (USDA, 2000g, 2001c, 2001d).

The committee was able to draw some general conclusions about the effectiveness of REE quality-assurance mechanisms and their outcomes.

## Peer Review

Internal peer reviews of individual projects are conducted annually and at project completion in all the REE agencies. Research programs receive periodic reviews, generally at 5-year intervals. The committee found that reviews tended to report intermediate outputs, publications, presentations, patents, cultivars, and breeds developed. Although such activities provide useful information, they are by their nature directed more toward quantity than quality of the effort. External peer review of project proposals and products is also occurring in the REE agencies.

### Intramural Funding

In response to previous studies critical of the lack of peer review of research proposals in the ARS system, major changes were introduced in 1999, as required by the 1998 AREERA (US Congress, 1998). These changes mandated periodic (usually 5-year) peer review of all research-project plans and creation of an Office of Scientific Quality Review (OSQR). Review by scientists outside ARS is involved in the process. Unsatisfactory research plans are not approved and must be rewritten; in some cases, the project must be terminated.[4] However, as of

---

[4]Offices of the area directors manage the postpanel activities of project plans receiving a "major revision" or "not feasible" action class. Projects falling into these categories may be dealt with in three ways (frequency of management action as of August 2002 is shown in parentheses): (1) prepare a revised project plan for a re-review by the panel within 3 months (95%), (2) completely rewrite the plan and have it peer-reviewed again by a new combination of reviewers about a year later (4%), (3) terminate the project (less than 1%). Re-review by the original panel is the most common action. A complete rewrite of the project plan and fresh peer review are most often the option chosen if the reason for the poor score was the absence of key scientific expertise (usually a result of a vacant position). The original panel reviewers are asked to provide a second review in such cases. Finally, projects may be terminated if there are extenuating personnel issues or an inability to correct the problems. Successful objectives from ARS projects that have been terminated because of overall poor peer reviews were transferred to a complementary research project. The scientists involved were either reassigned or placed under other personnel actions. Although grade level is not affected by the reassignment, a scientist may lose his or her status as a lead scientist on a project that is terminated, without a change in salary. The termination of projects is usually chosen only after a second review of the proposal results in a judgment of "not feasible" or "major revision required."

August 2002, OSQR reports that only five of several hundred projects have been terminated; none of the terminations has resulted in the removal of a scientist. The impact of an unsatisfactory research proposal on a scientist's performance evaluation is handled case by case, depending on the nature of the problem.

The program appears to have encouraged the development of improved research proposals, which does not necessarily guarantee that the quality of research in a given unit will be improved but suggests that the quality of the overall research program should increase with time. However, the low rate of project termination resulting from either proposal review or performance evaluation suggests limited accountability of the system. The major research costs (salary for ARS scientists and laboratory personnel) constitute fixed expenses for a unit regardless of whether specific research proposals have been approved, and the procedure does not represent competition or ranking among proposals for these funds. The national program leader for each research program category is responsible for priority-setting and quality assurance but does not have the requisite authority to move personnel or budget to various research centers to ensure those outcomes.

The committee considered results of reviews of six national programs that underwent peer review from February 2000 to August 2001 (Table 6-2). The most important evidence from Table 6-2 is that 23% of the program reviews asked for major revisions. Of the 94 review-panel members, 12 were employed by government agencies other than ARS, 69 were university faculty members, 12 were industrial or private consultants, and one was employed by ARS. The makeup of the review panels demonstrates that ARS programs are subjected to outside review.

External peer-review mechanisms are also in place in ERS, where all published materials are reviewed by experts for appropriateness for the category of publication. ERS's FY 2000 performance report states that published research meets peer-review standards for all five goals in every case (USDA, 2000e). ERS also has an internal peer-review process in place for research proposals.

At NASS, in the currently narrowly defined focus of NASS efforts (see Chapter 4), the quality-assurance processes are focused on the value of reports to particular end users. Customer satisfaction surveys and end-user meetings provide little information on the relative value of different types of estimates or information. In addition to end-user feedback, NASS quality assurance would also benefit from regular peer review of survey and estimation (forecast) methods by academic and other government-agency statisticians to ensure that they reflect the newest analytic techniques.

The committee considered REE's intramural research quality-assurance mechanisms against the mechanisms for quality assurance used in two other federal intramural research programs—those of the National Institutes of Health (NIH) and the Environmental Protection Agency (EPA). Intramural research at NIH has been subject to external scientific review since 1956. NIH's Office of

**TABLE 6-2** Results of February 2000–August 2001 Review of Six ARS National Programs

| National Program (NP) | Number of Projects Reviewed | Action Needed after Review | | | | |
|---|---|---|---|---|---|---|
| | | No Revision[a] | Minor Revision[b] | Moderate Revision[c] | Major Revision[d] | Not Feasible[e] |
| Manure and Byproduct Utilization (NP 201) | 21 | 1 | 10 | 7 | 2 | 1 |
| Food Safety (NP 108) | 62 | 1 | 21 | 21 | 18 | 1 |
| Soil Resources and Management (NP 206) | 33 | 8 | 14 | 6 | 5 | 0 |
| Plant Biological and Molecular Processes (NP 302) | 36 | 1 | 13 | 8 | 13 | 1 |
| Animal Health (NP 103) | 35 | 1 | 8 | 15 | 9 | 2 |
| Water Quality (NP 201) | 31 | 7 | 12 | 8 | 4 | 0 |
| Total | 218 | 19 | 78 | 65 | 51 | 5 |

[a] *No revision required.* No revision is required, but minor changes to project plan may be made.
[b] *Minor revision required.* Project plan is basically feasible as written but requires some revision to increase quality.
[c] *Moderate revision required.* Project plan is basically feasible as written but requires moderate revision of one or more objectives, perhaps involving changes in experimental approaches, to increase quality. Project plan may also need rewriting for greater clarity.
[d] *Major revision required.* Substantial revision of one or more objectives is necessary, but project plan should then be sound and feasible.
[e] *Not feasible.* Project plan has major flaws or deficiencies and cannot be simply revised to produce sound project. If project is not terminated, complete redesign and rewriting are required.

Source: ARS, Office of Scientific Quality Review, 2001.

Intramural Research reported to the committee that NIH has a rigorous, largely retrospective, review system in place in its intramural research programs. In contrast with the review of extramural grants, which mainly assesses the quality of proposed research, the work of all principal investigators is reviewed mainly in retrospective fashion, in which the research program is evaluated in toto for its overall goals, quality of research, and long-term objectives, based in specific criteria (Box 6-1) (USDHHS, 2002). In the case of new investigators, more emphasis is placed on future plans. Principal investigators are either tenured or on tenure track, designations conferred only after rigorous searches, peer review, selection processes, and internal and external reviews of research programs. The

### BOX 6-1
### Criteria for Review of NIH and ARS Intramural Research: A Comparison

**Criteria for Review of NIH Intramural Research**

*Significance*
Have the investigator's studies addressed important problems? Are the aims of the project(s) being achieved? Is scientific knowledge being advanced, and are the projects affecting the concepts or methods that drive this field?

*Approach*
In general are the approaches well conceived? When problem areas arose, were reasonable alternative tactics used?

*Innovation*
Do the projects use novel concepts, approaches, or methods? Are the aims original and innovative? Do the projects challenge existing paradigms or develop new methodologies or technologies?

*Environment*
Is the investigator taking advantage of the special features of the NIH intramural scientific environment or employing useful collaborative arrangements?

*Support*
Is the support the investigator received appropriate?

*Investigator Training*
Is the investigator appropriately trained and well suited to carry out the projects being pursued? Is the work proposed appropriate to the experience level of the principal investigator and other researchers (if any)?

*Productivity*
Considering the investigator's other responsibilities (e.g., service or administrative), how would you rate his/her overall research productivity?

*Mentoring*
Is the investigator providing appropriate training and mentoring for more junior investigators?

**Criteria for Review of ARS Research Proposals**

*Merit and Significance*
For this criterion, ARS is primarily interested in whether the problems to be solved or addressed fit within the National Program Action Plan to which the Project Plan is assigned. The National Program Action Plan has been developed with input from stakeholders, congressional mandates, customers, and ARS and non-ARS scientists. Other aspects of these criteria that should be addressed are: Will the successful completion of the project enhance knowledge of a scientifically important problem? Will the project lead to the development of new knowledge and technology? Are other data/studies relevant to this research effort? If applied research, of what value is the research to its customers?

*Adequacy of Approach and Procedures*
This evaluation criterion measures the scientific quality of the proposed research. Questions to be answered are: Are the hypotheses and/or plan of work well conceived? Are the experiments, analytical methods, and approaches and procedures appropriate and sufficient to accomplish the objectives? How could the approach or research procedures be improved?

*Probability of Successfully Accomplishing the Project Objectives*
The feasibility of the project is evaluated by this criterion. The panel will determine the probability of success in light of the investigator of project team's training, research experience, preliminary data if available, and past accomplishments, whether the objectives are both feasible and realistic within the stated time frame and with the resources proposed, and whether the investigators have an adequate knowledge of the literature as it relates to the proposed research.

Source: USDA (2000h); USDHHS (2002).

success rate for achieving tenure is about 65%, so 35% of the scientists who enter a 6-year tenure-track period do not compete successfully for scientific research resources (or leave for other reasons). Independent research support is provided to all principal investigators, and their progress is evaluated at least every 4 years by groups of outside experts, constituted as boards of scientific counselors. The emphasis of the outside reviews is mainly on past accomplishments, although future plans are presented to the review teams as well. Reviewers develop written recommendations to increase, decrease, or hold constant the resources assigned to principal investigators; these recommendations are acted on by the scientific program leaders. During the past 5-year period in one of NIH's major institutes, 7% of principal investigators lost all research support, and 25% of principal investigators had resources reduced. Resources of other principal investigators were increased or held at the same levels, leaving room for expansion of new initiatives and programs.

The committee also considered mechanisms for peer review of intramural research at EPA. Peer review occurs at multiple levels. For example, EPA's National Health and Environmental Effects Research Laboratory, in the Office of Research and Development, conducts reviews of each of nine divisions, reviewing research strategies and multiyear implementation plans, cross-cutting research programs, specific scientists undergoing promotion, and investigator-initiated research proposals. A variety of review processes are used, including ad hoc panel reviews, internal review by EPA experts, Federal Advisory Committee Act reviews, and reviews by ad hoc panels comprising a mixture of discipline-specific external experts and internal standing committees. All major scientific or technical work products also undergo review by internal and external experts before their release.

**FINDING: The committee commends ARS and ERS for establishing peer-review processes. The ARS peer-review process assists researchers in producing higher-quality proposals, which are a necessary, but not an exclusive, component of higher-quality research. However, the ARS peer-review system appears to reward excellent research performance adequately but may not adequately exclude poor research performance, given the noncompetitive (unranked) nature of the peer-review process, the extremely low rates of project termination, and the lack of impact of poor performance in proposal peer review on personnel grade level.**

**RECOMMENDATION 7: The REE intramural research system should strengthen quality control for poor research performance. Mechanisms used at other federal intramural research agencies, including the redirection of human or financial resources when quality is poor, could be implemented.**

The decision to terminate a project or a position should be made after a broad review of all aspects of a research program, including research inputs and outputs.

*Formula Funding*

CSREES uses review mechanisms to ensure the quality of a diverse portfolio of research, education, and extension programs with universities and other organizations. In some institutions, formula funds are built into base salary budgets of individual faculty; in others, the funds are distributed as specific research grants to individual faculty. The 1998 AREERA (US Congress, 1998) mandated mechanisms to ensure that each proposal undergoes merit or peer review to determine its quality before funding. Institutions eligible for formula funds are now required to document in their plans of work "a description of the merit and/or peer review process, . . . the selection of reviewers with expertise relevant to the effort, and appropriate scientific and technical standards" (USDA, 1998). The effectiveness of these policies in achieving improved accountability, however, remains unclear. Progress in formula-funded projects is assessed by CSREES through the receipt of interim and terminal CRIS reports.

*Competitive Grants*

In the CSREES competitive-grant programs, the National Research Institute (NRI), Initiative for Future Agriculture and Food Systems (IFAFS), and the Fund for Rural America (FRA) competitive programs have used a competitive merit-based, peer-review process to ensure quality. A National Research Council review of the NRI concluded that the quality of scientific work in this program was high (NRC, 2000).

*Special Grants*

CSREES special grants are added to the annual budget of the agency by congressional action. They are not subject to merit-based peer review and are directed to specific locations. CSREES can only ensure that special-grant programs represent the best quality in the institution or program by subjecting to peer review the output of special grants. The proposal or input of special grants can also be reviewed before the awarding of funds. In some cases, CSREES does not release the funds until the institution has hired a person with the necessary expertise or partnered with another institution that has the necessary expertise.

**FINDING: Several recent REE research programs (including the NRI, FRA, and IFAFS) are based on competitive peer-reviewed funding mechanisms, which make important contributions to ensuring the quality of the proposed research. Merit-based or peer-review mechanisms**

have improved the accountability of allocating formula and program funding to scientists. Peer review can be used as a mechanism for improving the quality of special grants.

## Quality of Research Conducted for Action Agencies

As discussed in Chapter 4, the REE agencies provide information to several action agencies, including the Animal and Plant Health Inspection Service, the Food and Nutrition Service, the Farm Services Agency, the Food Safety and Inspection Service, and the Natural Resources Conservation Service. In general, the committee's interviews with key stakeholders in client agencies in USDA indicated a high quality of research for short-term needs, although some agencies reported that some technologies developed were too complex to be user-friendly or were not relevant to the needs of the agency. Processes for tracking and reviewing REE research quality, responsiveness, and timeliness were cited as needing improvement, although action agencies did provide examples of ERS, ARS, NASS, and land-grant university responsiveness to data and research needs.

**FINDING: ARS, ERS, and land-grant university scientists are recognized for properly documented research that supports action agencies. There is no formal process for assessing the quality of REE support of action-agency needs, and there is no structure to track and review interactions between the agencies.**

## Performance Standards, Promotions, and Rewards

For government employees, the period of service in REE agencies is linked to salary step increases within and beyond (via the senior executive series) the Government Service (GS) grid. In addition, a system of rewards and incentives, including advancement in the grid, is based on a standardized evaluation using performance objectives or peer review. The system is a mechanism for ensuring the delivery of high-quality science.

REE agencies handle performance management differently. CSREES and NASS use panels to review staff performance. ARS and ERS use a peer-review system to determine promotions. In ARS, scientists are evaluated in relation to a performance plan and objectives that are established by the Administrative Office, with input from the ARS Office of Scientific Quality Review, under the Research Position Evaluation System (USDA, 2002b; see Box 6-2). The performance evaluation includes judgments on impact, research findings, publications, reporting, and planning capabilities. The process is clearly defined, and the review involves peer scientists outside the center of employment. Overall, performance evaluation of scientists for promotion in the ARS is an excellent program that fairly measures productivity and quality of scientific output. ERS instituted a similar peer-review system called the Economist Position Classification System in 1996 to determine position grade levels (see Box 6-2; USDA, 2000d).

*Internal and External Recognition of Research Quality*

The committee found evidence that REE scientists produce research of high quality. REE scientists have received national and international awards that show the high regard of their peers. Although CSREES did not provide the committee with data on university-level award recipients, Table 6-3 shows that among the

---

**BOX 6-2**
**Research Position Evaluation System at ARS and the Economist Position Classification System at ERS**

The Research Position Evaluation System (RPES) and the Economist Position Classification System at ERS (EPCS) provide for cyclic review of scientists to ensure classification accuracy. ARS and ERS have mandatory reviews every 3 years for GS levels 11–12, every 4 years for GS 13 positions, and every 5 years for GS 14–15. The objective of these reviews is to assign the grade level that best matches a person's qualifications.

RPES and EPCS are based on the "person-in-the-job" concept: research scientists and economists have open-ended promotion potential based on their personal research and leadership accomplishments, which can change the complexity and responsibility of their positions. EPCS removes the requirement that an employee go into management as he or she progresses. Excellent researchers can advance without moving into a management track.

Both systems ensure scientific discipline (peer-group) diversity in representation on the evaluation panels to allow for greater objectivity in decision-making. Peer scientists from two peer groups serve on each review panel.

EPCS applies to all economists and other social scientists in nonsupervisory, nonmanagement positions that involve research and analysis, program planning and administration, and consultant and advisory activities. RPES applies only to ARS category 1 research positions (permanent, independent scientists). Other professional scientific positions are evaluated by application of appropriate US Office of Personnel Management classification standards.

Source: USDA (2002b, 2000d).

**TABLE 6-3** 1999–2000 Intramural (ARS and ERS) Recipients of Major Awards Sponsored by External Organizations

| Sponsor | Agency | Number |
| --- | --- | --- |
| National Society of Professional Engineers | ARS | 1 |
| American Society for Horticultural Science | ARS | 1 |
| American Society for the Nutritional Research Sciences and the Mead Johnson Nutritional Companies | ARS | 1 |
| Trece Company | ARS | 2 |
| Entomological Society of America, Southeastern Branch | ARS | 2 |
| Commonwealth Scientific Industrial Research Organization | ARS | 1 |
| Department of Geological and Environmental Sciences, Stanford University | ARS | 1 |
| Alexander S. Onassis Public Benefits Foundation | ARS | 1 |
| American Society of Agronomy | ARS | 1 |
| Crop Science Society of America | ARS | 1 |
| American Dietetic Association | ARS | 1 |
| Pittsburgh Conference on Analytical Chemistry and Applied Spectroscopy | ARS | 1 |
| Washington State University, Pfizer Animal Health | ARS | 1 |
| American Oil Chemists Society | ARS | 1 |
| American Peanut Research and Education Society | ARS | 1 |
| American Veterinary Medical Association | ARS | 1 |
| American Society of Civil Engineers | ARS | 1 |
| Agronomic Science Foundation, Soil Science Society of America | ARS | 1 |
| American Agricultural Economics Association | ERS | 2 |
| Department of Energy | ARS | 3 |
| Rhone-Poulenc Award and Organization of Nematologists of Latin America | ARS | 1 |
| Mississippi Chapter of the American Society of Agronomy | ARS | 1 |
| Strategic Environmental Research and Development Program (Department of Defense, Department of Energy, the Environmental Protection Agency) | ARS | 1 |
| National Academy of Sciences | ARS | 1 |
| American Society of Agricultural Engineers | ARS | 3 |
| National Science and Technology Council (White House Presidential Early Career Awards for Scientists and Engineers) | ARS | 2 |
| North American Blueberry Council | ARS | 1 |
| Federal Laboratory Consortium | ARS | 25 |
| National Pork Producing Council | ARS | 3 |
| American Chemical Society and Royal Society of Chemistry | ARS | 1 |
| American Dairy Science Association | ARS | 1 |
| George Washington University | ARS | 1 |
| Women in Science and Engineering | ARS | 1 |

Source: Performance and Awards Staff, Human Resources Division, ARS (2001); USDA (2001a).

intramural research agencies there are a large number of recipients of awards. In addition, 10 of the 75 US members of the National Academy of Sciences in the two sections of primary interest to agriculture spent all or most of their careers with ARS, and most of the other members of the two sections are affiliated with land-grant universities (NAS, 2001). REE scientists have also been recipients of many internal awards, including the USDA secretary's Honor Awards. In 2000, seven were awarded to ERS scientists and 68 to ARS scientists. Incentives for performing high-quality research also exist within agencies. ERS reported that it uses cash awards to reward high-quality work.

## IMPACT ASSESSMENT

Agricultural research has primary, intermediate, and longer-term impacts, which are frequently summarized as science indicators and quantitative impacts. Agricultural research also has impacts measurable in net social surplus,[5] which have been summarized largely as social rates of return. Although quantitative data are not as systematized as economic rates of return, environmental, social, nutritional, health, and other indicators can be used to assess impact. A particular program or institution can have an impact on scientific progress as reflected in such changes as the adoption of new technology. Mission-oriented research programs must be further assessed on the basis of research outcomes consistent with the mission. A high-impact program may include projects that fulfill an institution's mission, such as providing information to other government agencies (for example, USDA's action agencies) and the Office of Management and Budget—research that may yield low rates of scientific publication.

In this section, the committee reviews the impact of REE research by using a variety of metrics focused on some dimension of the real output or payoff of research, including publications, citation frequency, patenting, longer-term quantitative measures, and social rate of return. The committee also reviews REE's internal mechanisms for tracking and monitoring its impact and provides recommendations for changes in them.

### Primary, Intermediate, and Longer-Term Impacts

Research outputs used as inputs to others' scientific work, publications, and patents are frequently called "leading science indicators" to show primary and intermediate impacts of research.

### Publications

Publications are a primary measure of research output. About 10,000 scientists are engaged in public[6] agricultural research in the United States (USDA,

---

[5]Net social surplus is the net change in producer and consumer surplus from shifts in the supply and/or demand curves. In the case of new agricultural technology, the increased productivity results in a positive change in net social surplus, because the same output can now be produced with fewer inputs.

[6]"Public" refers to the combination of federal research institutions and federally supported state institutions.

2000c). In 2000, the REE agencies produced 2,035 scientist-years of research, and the federally supported state research institutions (state agricultural experiment stations [SAESs] and land-grant universities) 7,332 scientist-years of research (USDA, 2000c). ARS scientists submitted over 9,300 papers to peer-reviewed journals in 1998–2001 and ranked first internationally in papers published in 1991–2001 in agricultural sciences and plant and animal sciences and second in papers published in environment and ecology (Table 6-4; ISI, 2001). Similar data are not available for research supported by CSREES and ERS.

The scientific community often uses citation frequency as an intermediate measure of research impact. In the three fields—environment and ecology, plant and animal sciences, and agricultural sciences—the impact of ARS publications was similar to, but somewhat lower than, that of publications from land-grant universities (Table 6-4). In plant and animal sciences, 15 US land-grant universities rank internationally in the top 20 (data not shown; ISI, 2001).

**TABLE 6-4** World Institutional Rankings in Select Fields, by Total Citations, 1991–2001

| Rank | Institution | Number of Papers | Number of Citations | Number of Citations per Paper |
|---|---|---|---|---|
| **Agricultural Sciences** | | | | |
| 1 | USDA ARS | 5,603 | 28,986 | 5.17 |
| 2 | Institut National de la Recherche Agronomique (INRA) (France) | 2,698 | 14,919 | 5.53 |
| 3 | University of California Davis | 1,463 | 9,868 | 6.75 |
| 4 | Commonwealth Scientific and Industrial Research Organisation (CSIRO) (Australia) | 1,461 | 9,049 | 6.19 |
| 5 | University of Wisconsin | 1,322 | 9,019 | 6.82 |
| **Plant and Animal Sciences** | | | | |
| 1 | USDA ARS | 9,908 | 63,201 | 6.38 |
| 2 | University of California Davis | 5,584 | 39,186 | 7.02 |
| 3 | Cornell University | 4,451 | 36,966 | 8.31 |
| 4 | INRA (France) | 5,833 | 35,467 | 6.08 |
| 5 | University of Wisconsin | 3,510 | 27,442 | 7.82 |
| **Environment and Ecology** | | | | |
| 1 | US EPA | 2,049 | 18,875 | 9.21 |
| 2 | USDA ARS | 2,450 | 18,806 | 7.68 |
| 3 | CSIRO (Australia) | 1,523 | 14,365 | 9.43 |
| 4 | University of California Berkeley | 1,170 | 12,392 | 10.59 |
| 5 | University of Minnesota | 1,055 | 12,218 | 11.58 |

Source: Adapted from ISI Essential Science Indicators, 2001. ScienceWatch: Trends and Performance in Basic Research. July/August, 2001. ISI Essential Science Indicators, 1991–2001.

## Patents

Patents are the products of invention and can indicate that research results are being accepted and used in the US economy. USDA received an average of 57 patents per year from 1996 to 2000 (see Chapter 5, Table 5-2).

## Social Rate of Return

The social rate of return on public research expenditures is the economist's preferred tool for assessing impact; it provides a bottom-line estimate in the form of a rate of return. The return can be compared with the social opportunity cost of public funds. Sometimes it is difficult, however, to obtain the time profile of net social benefits of research studies.[7] Society will be best served if it invests its research resources into projects that yield a sizable positive social marginal rate of return.

Societal benefits in the form of rates of return from agricultural research have been extensively reviewed (see Tables 6-5 and 6-6). Rates of return are high but vary widely. Evenson's summary of rates of return on public aggregate[8] agricultural research investments shows a median real social rate of return from 126 research studies of 45%; 66% were in the range of 21–80%. The research studies summarized by Evenson generally estimate the market value of improvements in agricultural productivity (this results in lower consumer prices or lower costs for producers over time) and then compare that value with the costs of investments in agricultural research. Alston et al. (2000) also published a formal and extensive analysis (292 studies, 1,886 rate-of-return observations), which econometrically accounts for the observed variation in the rates of return. That study showed that the mean of the measured rates of return to research (averaging over 1,144 observations) was 99.6% with a mode of 46%. Both meta-studies support the finding that investments in agricultural research have generated high and sustained returns to society.

---

[7]When individual projects are evaluated, it generally takes several years after a study is completed before the social cost-benefit calculations can be made.

[8]Public denotes a combination of state and federal, and the aggregate value is only a proxy for REE research. Assessment of the impact of research supported by USDA formula and competitive funds at universities is difficult. Individual universities decide how the formula funds are to be allocated among competing needs, including research, education, and outreach. Federal funds from REE are most often combined with other funds (from state, private, and other federal agencies), and attribution of the individual contributions is nearly impossible with today's accounting procedures. Returns from aggregate values seem most relevant here because they include successful and failed projects and provide a better picture of the whole enterprise than rates of return for particular studies. In the case of the SAES, in the period 1927–1980, when these rates of return were calculated, most funding came from state or federal sources, not from the private sector (see footnote 11).

**TABLE 6-5** Internal Rates of Return from US Public-Sector Agricultural Research

| Study | Method | Period | Internal Rate of Return,[a] % | Region |
|---|---|---|---|---|
| Peterson and Fitzharris (1977) | Project evaluation | 1937–1942 | 50 | |
| Peterson and Fitzharris (1977) | Project evaluation | 1947–1952 | 51 | |
| Peterson and Fitzharris (1977) | Project evaluation | 1957–1962 | 49 | |
| Peterson and Fitzharris (1977) | Project evaluation | 1957–1972 | 34 | |
| Norton and Paczkowski (1993) | Project evaluation | 1949–1979 | 58 | VA |
| Norton and Paczkowski (1993) | Project evaluation | 1949–1989 | 58 | VA |
| Griliches (1964) | Statistical method | 1949–1959 | 25–40 | |
| Latimer (1964) | Statistical method | 1949–1959 | n.s.[b] | |
| Evenson (1968) | Statistical method | 1949–1959 | 47 | |
| Cline (1975) | Statistical method | 1939–1948 | 41–50 | |
| Bredahl and Peterson (1976) | Statistical method | 1937–1942 | 56 | |
| Bredahl and Peterson (1976) | Statistical method | 1947–1957 | 51 | |
| Bredahl and Peterson (1976) | Statistical method | 1957–1962 | 49 | |
| Bredahl and Peterson (1976) | Statistical method | 1967–1972 | 34 | |
| Lu et al. (1979) | Statistical method | 1938–1972 | 24–31 | |
| Lu et al. (1979) | Statistical method | 1939–1972 | 23–30 | |
| Evenson (1979) | Statistical method | 1868–1926 | 65 | |
| Evenson (1979) | Statistical method | 1927–1950 | 95 | |
| Evenson (1979) | Statistical method | 1948–1971 | 130 | South |
| Evenson (1979) | Statistical method | 1948–1971 | 93 | North |
| Evenson (1979) | Statistical method | 1948–1971 | 95 | West |
| Knutson and Tweeten (1979) | Statistical method | 1949–1972 | 28–47 | |
| White et al. (1978) | Statistical method | 1929–1977 | 28–37 | |
| Davis (1979) | Statistical method | 1949–1959 | 66–100 | |
| Davis and Peterson (1981) | Statistical method | 1949 | 100 | |
| Davis and Peterson (1981) | Statistical method | 1954 | 79 | |
| Davis and Peterson (1981) | Statistical method | 1959 | 66 | |
| Davis and Peterson (1981) | Statistical method | 1964, 1969, 1974 | 37 | |
| Welch and Evenson (1989) | Statistical method | 1969 | 55 | |
| White and Havlicek (1982) | Statistical method | 1943–1977 | 7–36 | |
| Braha and Tweeten (1986) | Statistical method | 1959–1982 | 47 | |
| Evenson (1989) | Statistical method | 1950–1982 | 43 | |
| Alston et al. (1998a) | Statistical method | | 17–31 | |
| Chavas and Cox (1992) | Statistical method | | 28 | |
| Makki et al. (1996) | Statistical method | 1930–1990 | 27 | |
| Makki and Tweeten (1993) | Statistical method | 1930–1990 | 93 | |
| Oehmke (1996) | Statistical method | Pre–1930 | Negative | |
| Oehmke (1996) | Statistical method | 1930–1990 | 11.6 | |
| Yee (1992) | Statistical method | 1931–1985 | 49–58 | |
| Norton et al. (1992) | Statistical method | 1987 | 30 | |

[a]The internal rate of return is a discount rate at which the present value of a series of investments is equal to the present value of the returns on those investments.
[b]n.s.= not significant.

Source: Adapted from Evenson (2001). Economic Impacts of Agricultural Research and Extension in Agricultural Economics Volume 1a Agricultural Production, B.L. Gardner and G.C. Rausser, eds. Amsterdam: Elsevier.

**TABLE 6-6** Internal Rates of Return from US Extension

| Study | Period | Extension Variable | Internal Rate of Return,[a] % |
|---|---|---|---|
| Huffman (1974)[b] | 1959–1974 | Extension staff/farm | 16 |
| Huffman (1976)[b] | 1964 | Staff days/farm | 110 |
| Evenson (1979)[b] | 1971 | Expenditure/region | 100+ |
| Huffman (1981)[b] | 1979 | Extension days/county | 110 |
| Evenson (1994)[b] | 1950–1972 | Expenditure/state | Crops 101 |
| Evenson (1994)[b] | 1950–1972 | Expenditure/state | Livestock 89 |
| Evenson (1994)[b] | 1950–1972 | Expenditure/state | All 82 |
| Huffman and Evenson (1993)[c] | 1950–1982 | | Crops 40.1 |
| Huffman and Evenson (1993)[c] | 1950–1982 | | Livestock (negative) |
| Huffman and Evenson (1993)[c] | 1950–1982 | | All 20.1 |

[a]The internal rate of return is a discount rate at which the present value of a series of investments is equal to the present value of the returns on those investments.
[b]Evenson, R. 2001. Economic impacts of agricultural research and extension. In Agricultural Economics Volume 1a. Agricultural Production. B.L. Gardner and G.C. Rausser, eds. Amsterdam: Elsevier. p. 593.
[c]Huffman, W.E., and R.E. Evenson. 1993. Science for Agriculture: A Long-Term Perspective. Ames, IA: Iowa State University Press, p. 245.

Other general observations over the 1950–1982 period include a much higher rate of return from preinvention or pretechnology research than from applied research (Evenson, 2001), a higher rate of return from applied crop research than from applied livestock research (Huffman and Evenson, 1993), and a lower rate of return as the share of public agricultural research funding by federal contracts, grants, and cooperative agreements increased (Huffman and Just, 1994).

High rates of return are consistent with growth in total factor productivity.[9] Jorgenson and Stiroh (2000) show that the US agricultural sector ranks third among 37 US industries in aggregate total factor productivity growth over 1958–1996 at 1.2%. New total factor productivity estimates by ERS for US agriculture show an average annual rate of 2% for 1950–1998. Evenson (2001) describes how a continuous investment in public agricultural research of 1% of output per year will contribute 0.76% per year to total factor productivity growth when the real rate of return is 40%.

---

[9]A total-productivity or multifactor productivity index measures the quantity of output produced compared with the quantity or cost of all the measurable inputs used to produce it (Ahearn et al., 1998). Growth in productivity indicates the growth in output unaccounted for by changes in measured inputs and is typically ascribed to investments in R&D, education, infrastructure, and economies arising from increasing the scale of production.

The high rates of return from public agricultural research summarized by Evenson (2001) pertain to investments made in public[10] agricultural research from 1927 to 1980. Alston et al. (2000), using econometric evidence after accounting for factors that cause the reported rates of return to vary among studies, established that the rate of return over the last 2 decades has not declined.

It is important to discuss the inherent limitations of using social rate of return as an impact indicator. For example, the estimates of rate of return do not account for possible off-farm environmental costs or benefits (Jorgenson and Stiroh, 2000). In addition, rate of return does not account for the full economic costs and benefits for farms of different sizes and for agriculture-dependent communities (for example, new technologies may enhance productivity of larger farms relatively but contribute to the decline of smaller farms and the rural communities that were once sustained by these farms) (Swanson, 1988; US Congress OTA, 1986).

**FINDING: The social rate of return on past public agricultural research investments in the period 1950–1982 has been very high. The rate of return over the last 2 decades has not declined. Social rate of return has limitations in accounting for full environmental and social costs of research.**

### Environmental, Economic, and Health Outcomes

Aside from the long-run benefits of productivity enhancement measured through social rate of return, documented quantitative examples of the impact of REE research on other outcomes are scarce. Box 6-3 provides several illustrative examples of successful research that provided benefits to the environment, health, or safety that are difficult to quantify on a national scale. As the examples show, successful impact of specific research projects may be obvious, but difficult to summarize in comparable measures. That is particularly true for environmental benefits, human health benefits, or potential social benefits. Although the research frontiers identified in this report will result in such benefits, they may be difficult to quantify or compare among research goals and projects.

---

[10]The private-sector share of the total support for SAESs was 7% in 1960, 9.2% in 1980, and 13.2% in 1990. Before 1960, disaggregated data for the private-sector contribution are not available, but we know that the sum of the private-industry contribution and other federal-government (non-USDA) resources was not greater than 14.8% in 1900, 29.8% in 1930, and 22.6% in 1940. Thus, USDA appropriations and state-government appropriations predominated as the source of support of the SAESs (85% in 1900, 70.2% in 1920, 77.4% in 1940, 79.8% in 1960, and 72.5% in 1980) during the period in which these rates of return were calculated (Huffman and Evenson, 1993).

## BOX 6-3
## Examples of REE Research Impacts

### Improvement in Water-Quality Practices

The USDA Water-Quality Initiative has led to the adoption of practices that substantially reduced applications of pesticides, nitrogen, and phosphorus on over 500,000 acres of midwestern farmland (Amerman et al., 2001). The program required the coordinated efforts of several USDA agencies, the Department of the Interior, the Department of Commerce, and EPA.

### Eradication of the Screwworm

An example of the impact of very successful USDA research, one of the greatest entomologic success stories of all time, is eradication of the screwworm. Obnoxious and destructive, the screwworm is the only insect known to consume the living flesh of warm-blooded animals and has caused much suffering and losses in livestock, wildlife, and human populations the world over. ARS scientists reasoned that sterilized screwworm males released into infested areas would mate with fertile females and lead to a reduction in screwworm population. The sterile-fly release program led to elimination of the screwworm from the United States, Mexico, and several countries in Central America. Billions of dollars in savings and reduction in suffering occurred from this program, and the benefits have accrued over several decades (USDA, 1992).

### Economic Benefits of Investments in Potato Research

Production of potatoes is concentrated in the Pacific Northwest, which produces about 66% of the nation's potatoes; Idaho accounts for 30%. Annual public research investments in potato research by the top 21 producing states averaged over $26.7 million in 1987–1991. SAES researchers at the University of Idaho conducted a cost-benefit analysis of public potato research over 1967–1990, using data on 21 states grouped into six regions that had homogeneous geography, climate, production methods, and type of potato produced. The estimated economic benefit of public investment in potato research takes account of spillover benefits of research results across regions. The study showed that the rate of return at the national level on investment in public potato research, accounting for interstate spillover effects, was 79%. The average share of benefits accruing to states originating the research was 31 to 69% of the benefits accruing to other potato-producing states through spillover between regions. The sizable spillover benefits from originating-state public potato research suggest that interstate coordination of potato research is important for good public science-policy decision-making (Araji et al., 1996).

*continued*

> **BOX 6-3 Continued**
>
> **Economic and Environmental Impacts of Insect-Growth Regulators**
> A research project tracked the adoption and diffusion of insect-growth regulators (IGRs) on Arizona cotton production and their economic and environmental ramifications. In 1995, whiteflies in Arizona exhibited resistance to all the commonly used insecticides. In some areas, growers made 8 to 12 insecticide applications with costs of $200–$300 per acre (costing the state $57–86 million per year). Despite high pest-control investments, growers received price discounts as high as about 7–8% of gross revenues. In 1995, the University of Arizona, ARS, the Arizona Cotton Growers Association, and Cotton Incorporated undertook collaborative public–private research to gain EPA Section 18 exemptions to use IGRs to control whiteflies. As a result, EPA granted exemption in 1996. Through the use of geographic information systems, cotton-acreage data were overlaid on pesticide-use data to construct a valuable new database on insecticide use intensity. The database has been used to trace IGR diffusion patterns, to explain the adoption of IGRs in the state, and to estimate economic benefits of the new technology in terms of reduced grower costs and environmental benefits in terms of reduced overall insecticide use (Frisvold et al., 2002).
>
> **Health Impacts of Folic Acid Research**
> Ingestion of folic acid reduces blood plasma concentration of homocysteine, a cardiovascular risk factor when present in high amounts. ARS-funded nutrition research played a lead role in doubling the recommended daily allowance of folic acid. The research also showed a relationship between plasma homocysteine and carotid-artery stenosis, coronary avascular and total mortality, incidence of stroke, and increased risk of dementia and Alzheimer's disease; and it demonstrated the vital roles of sufficient folic acid in the human diet (Bostom et al., 1999a, 1999b; Selhub et al., 1995, 1998; Seshadri et al., 2002).

## Monitoring and Communicating Impact

Mechanisms for tracking research investments and impacts are important both for informing internal decision-making on future research investments and for communicating with and improving accountability to the public. The committee considered a variety of mechanisms used by REE to track and communicate its performance and impact. These are considered in detail in Appendix G.

This report has identified a need to broaden the REE agency focus on needs of and impacts on new, nontraditional stakeholders and to be more strategic in setting priorities for investments to meet national goals (Chapters 1 and 4). To achieve these goals, REE will need to improve and expand existing performance-based systems for monitoring success. More-effective and more user-friendly tracking systems will contribute to improved self-evaluation and reporting of progress to groups outside REE. Electronic media are an increasingly critical and strategic means for communicating impacts and research results to the general public and should be a focal point for development and expansion. ERS's recent redesign of its Web site to be more accessible and understandable to the public—through provision of links to nontechnical summaries, technical abstracts, data, and publications—is an excellent model for how positive change could occur in this regard throughout the REE mission area.

**RECOMMENDATION 8: REE agencies should develop and adopt ways of measuring the national, long-term impacts of their research on the environment, human health, and communities. The tools should include measures and indicators that are influenced by agricultural research or that can be attributed to research outcomes, including how research supports the needs of action agencies. REE should strive to achieve greater transparency in communicating these impacts through timely electronic publishing of peer-reviewed results and through greater efforts to interpret these results for a general audience.**

The committee envisions that monitoring capability and development of indicators would occur in parallel at two levels. First, monitoring capability could be developed to show how REE research has changed in focus, relevance, quality, leadership, and accountability (NRC, 1999a). For example, REE could track progress toward meeting national goals by requiring that each major research program or initiative establish performance objectives and measures for evaluating progress toward meeting national objectives, on the basis of some assessment of adoption and/or implementation of research findings in practical applications and possibly of how such adoption led to beneficial changes. In addition, REE could track performance by keeping a comprehensive account of where research funded or conducted by REE has made a critical contribution to advancing understanding, policies, and practices in each of the research frontiers recommended by this study.

A second level of monitoring capability could be developed to show how food, agricultural, natural, and human systems are changing. Because changes in such indicators cannot be directly attributed to research, this class of indicators cannot provide a direct measure of research performance, but it can be used to help target future research directions. Such indicators might include nutritional indicators, such as the healthy eating index measuring overall nutritional quality of the American diet (Kennedy et al., 1999), and ecologic indicators, including nutrient runoff and soil organic matter (NRC, 2000). Indicators should be selected after a set of defining criteria has been established, which might include a well-understood conceptual basis, reliability, applicability on clear temporal and spatial scales, accuracy, sensitivity, precision, robustness, skill and data requirements, data-quality requirements, archiving capability, international compatibility, costs, benefits, and cost effectiveness (NRC, 2000). As discussed in Chapter 3, systematic research on the environmental, social, and community impacts of REE research would inform this process.

## SUMMARY

This chapter has considered quality-assurance and impact-assurance processes and their outcomes in the REE agencies and has provided recommendations for improving the effectiveness of these processes. The use of peer review as a quality-assurance mechanism for research inputs and outputs was discussed with regard to intramural research, formula-funded research, special grants, and competitive grants. REE staff-performance evaluation systems and reward and incentive programs were also considered as mechanisms for ensuring the delivery of high-quality science. In general, REE has strong and evolving quality-assurance mechanisms in place, and REE scientists produce research that is of high quality. However, human or financial resources should be redirected when research is of poor quality in intramural research.

Primary, intermediate, and longer-term impacts of agricultural research were described. Much progress has been made in documenting research impact in the traditional dimensions associated with improved productivity. Because in the future more REE research will be directed at providing new kinds of benefits, monitoring of research impact will require new outcome measures. The importance of monitoring, measuring, and communicating research investments and their impacts was discussed, and changes in monitoring capability, in development of indicators, and in communication of results were recommended.

## REFERENCES

Ahearn, M., J. Yee, E. Ball, and R. Nehring. 1998. Agricultural Productivity in the United States. Agricultural Information Bulletin, No. 740. Washington, DC: Economic Research Service, US Department of Agriculture. Available online at *http://www.ers.usda.gov/publications/aib740/*.

Alston, J.M., M.C. Marra, P.G. Pardey, and T.J. Wyatt. 2000. Research returns redux: A meta-analysis of the returns to agricultural R&D. Australian Journal of Agricultural and Resource Economics 44(2):185–215.

Amerman, R.C., G. Larson, and M. O'Neill. 2001. USDA Water Quality Initiative. Presentation to National Research Council Committee on Opportunities in Agriculture, Public Workshop, May 22–23. Washington, DC.

Araji, A.A., F.C. White, and J.F. Guenthner. 1996. Returns from Potato Research: Accounting for State and Regional Effects. Agricultural Experiment Station, Research Bulletin 152. Moscow, ID: University of Idaho.

Arndt, M. 2002. 3M: A lab for growth? Business Week, Jan. 21.

Bostom, A.G., I.H. Rosenberg, H. Silbershatz, P.F. Jacques, J. Selhub, R.B. D'Agostino, P.W.F. Wilson, and P.A. Wolf. 1999a. Nonfasting plasma total homocysteine levels and stroke incidence in elderly persons: The Framingham study. Annals of Internal Medicine 131:352–355.

Bostom, A.G., H. Silbershatz, I.H. Rosenberg, J. Selhub, R.B. D'Agostino, P.A. Wolf, P.F. Jacques, and P.W.F. Wilson. 1999b. Nonfasting plasma total homocysteine levels and all-cause and cardiovascular disease mortality in elderly Framingham men and women. Archives of Internal Medicine 159:1077–1080.

Evenson, R.E. 2001. Economic Impacts of Agricultural Research and Extension. Pp. 575–628 in Handbook of Agricultural Economics, Vol. 1: Agricultural Production, B.L. Gardner and G.C. Rausser, eds. Amsterdam: Elsevier.

Frisvold, G., G.K. Agnew, and P. Baker. 2002. Effects of Insect Growth Regulators on Insecticide Use and Costs in Arizona Cotton. Proceedings of the Beltwide Cotton Conferences 1. Memphis, TN: National Cotton Council.

Huffman, W.E., and R.E. Evenson. 1993. Science for Agriculture: A Long-Term Perspective. Ames, IA: Iowa State University Press.

Huffman, W.E., and R.E. Just. 1994. Funding, structure, and management of public agricultural research in the United States. American Journal of Agricultural Economics 76(November):744–759.

ISI (Institute for Scientific Information) 2001. ScienceWatch: Trends and Performance in Basic Research. July/August. ISI Essential Science Indicators, 1991–2001. Philadelphia, PA: Institute for Scientific Information.

Jorgenson, D.W., and K.J. Stiroh. 2000. US economic growth at the industry level. American Economic Review 90(May):161–167.

Kennedy, E., S.A. Bowman, M. Lino, S.A. Gerrior, and P.P. Basiotis. 1999. Diet quality of Americans: Health eating index. Chapter 5 in America's Eating Habits: Changes and Consequences, E. Frazao, ed. Agriculture Information Bulletin No. 750. May. Washington, DC: Economic Research Service, US Department of Agriculture.

NAS (National Academy of Sciences). 2001. Membership Directory. July 2001. National Academy of Sciences of the United States of America.

NRC (National Research Council). 1999a. Evaluating Federal Research Programs: Research and the Government Performance and Results Act. Washington, DC: National Academy Press.

NRC (National Research Council). 1999b. Sowing the Seeds of Change: Informing Public Policy in the Economic Research Service. Washington, DC: National Academy Press.

NRC (National Research Council). 2000. National Research Initiative: A Vital Competitive Grants Program in Food, Fiber, and Natural-Resources Research. Washington, DC: National Academy Press.

Selhub, J., P.F. Jacques, A.G. Bostom, R.B. D'Agostino, P.W.F. Wilson, A.J. Belanger, D.H. O'Leary, P.A. Wolf, E.J. Schaefer, and I.H. Rosenberg. 1995. Association between plasma homocysteine concentrations and extracranial carotid-artery stenosis. New England Journal of Medicine 332:286–291.

Selhub, J., P.F. Jacques, P.W.F. Wilson, D. Rush, and I.H. Rosenberg. 1998. Vitamin status and intake as primary determinants of homocysteinemia in an elderly population. Journal of the American Medical Association 270(20):2693–2698.

Seshadri, S., A. Beiser, J. Selhub, P.F. Jacques, I.H. Rosenberg, R.B. D'Agostino, P.W.F. Wilson, and P.A. Wolf. 2002. Plasma homocysteine as a risk factor for dementia and Alzheimer's disease. New England Journal of Medicine 346(7):76–73.

Swanson, L. 1988. Agriculture and Community Change in the US: The Congressional Research Reports. Boulder, CO: Westview Press.

US Congress. 1998. P.L. (Public Law) 105–185. Agricultural Research, Extension, and Education Reform Act of 1998.

US Congress, OTA (Office of Technology Assessment). 1986. Technology, Public Policy, and the Changing Structure of American Agriculture. OTA F-285. Washington, DC: US Government Printing Office.

USDA (US Department of Agriculture). 1992. Subduing the Screwworm. Agricultural Research. July. Washington, DC: Agricultural Research Service, US Department of Agriculture.

USDA (US Department of Agriculture). 1998. Guidelines for Peer and Merit Reviews. Washington, DC: Cooperative State Research, Education, and Extension Service, US Department of Agriculture.

USDA (US Department of Agriculture). 1999. Agricultural Research Service Strategic Plan: Working Document 1997–2002. Washington, DC: Agricultural Research Service, US Department of Agriculture. Available online at http://www.nps.ars.usda.gov/mgmt/stratpln/1999/background.cfm.

USDA (US Department of Agriculture). 2000a. Agricultural Research Service FY 2000 and 2001 Annual Performance Plans. Washington, DC: Agricultural Research Service, US Department of Agriculture.

USDA (US Department of Agriculture). 2000b. Cooperative State Research, Education, and Extension Service FY 2000 and 2001 Annual Performance Plan. Washington, DC: Cooperative State Research, Education, and Extension Service, US Department of Agriculture.

USDA (US Department of Agriculture). 2000c. Current Research Information System (CRIS) Funding Summaries, Table A, FY 2000. Washington, DC: US Department of Agriculture. Available online at http://cristel.csrees.usda.gov/star/00tablea.pdf.

USDA (US Department of Agriculture). 2000d. Economist Position Classification System. December 2000. Washington, DC: Economic Research Service, US Department of Agriculture.

USDA (US Department of Agriculture). 2000e. Economic Research Service FY 2000 Annual Performance Report. Washington, DC: Economic Research Service, US Department of Agriculture. Available online at http://www.ers.usda.gov/AboutERS/ersannualperformance.pdf.

USDA (US Department of Agriculture). 2000f. Economic Research Service Strategic Plan, 2000–2005. October 11. Washington, DC: Economic Research Service, US Department of Agriculture. Available online at http://www.ers.usda.gov/AboutERS/ersstrategicplan.pdf.

USDA (US Department of Agriculture). 2000g. GPRA Strategic Plan. Washington, DC: National Agricultural Statistics Service, US Department of Agriculture. Available online at http://www.usda.gov/nass/nassinfo/strat-2005.pdf.

USDA (US Department of Agriculture). 2000h. Peer Review of ARS Research Project Plans. Office of Scientific Quality Review, Agricultural Research Service. November 22. Washington, DC: Agricultural Research Service, US Department of Agriculture. Available online at http://www.ars.usda.gov/osqr/OAManual.pdf.

USDA (US Department of Agriculture). 2001a. Data submitted to the National Research Council Committee on Opportunities in Agriculture. Washington, DC: Economic Research Service, US Department of Agriculture.

USDA (US Department of Agriculture). 2001b. Economic Research Service FY 2002 Annual Performance Plan and Revised Plan for FY 2001 (July 2001). Washington, DC: Economic Research Service, US Department of Agriculture. Available online at http://www.ers.usda.gov/AboutERS/ersperformance_plan.pdf.

USDA (US Department of Agriculture). 2001c. National Agricultural Statistics Service FY 2002 and Revised FY 2001 Annual Performance Plans. Washington, DC: National Agricultural Statistics Service, US Department of Agriculture. Available online at http://www.usda.gov/nass/nassinfo/nass-app-02-01.pdf.

USDA (US Department of Agriculture). 2001d. National Agricultural Statistics Service. FY 2000 Annual Program Performance Report. Washington, DC: National Agricultural Statistics Service, US Department of Agriculture. Available online at http://www.usda.gov/ocfo/ar2000/aprpdf/arnass.pdf.

USDA (US Department of Agriculture). 2002a. Information Quality Guidelines. Washington, DC: National Agricultural Statistics Service, US Department of Agriculture. Available online at *http://www.usda.gov/nass/nassinfo/infoguide.htm.*

USDA (US Department of Agriculture). 2002b. Research Position Evaluation System. Washington, DC: Agricultural Research Service, US Department of Agriculture. Available online at *http://www.afm.ars.usda.gov/rpes/.*

USDHHS (US Department of Health and Human Services). 2002. Orientation Guidelines for Boards of Scientific Counselors. Office of the Director, National Institutes of Health. Washington, DC: National Institutes of Health, US Department of Health and Human Services.

# 7

# REE Capacity

This chapter considers four major dimensions of the capacity of the US Department of Agriculture (USDA) Research, Education, and Economics (REE) mission area: organizational capacity, human capacity, information capacity, and infrastructure capacity. Those capacities provide the critical foundation for the production of high-quality research.

## ORGANIZATIONAL CAPACITY

Like any relatively large organization, the REE agencies function within a set of interrelated systems designed to perform a number of tasks in an increasingly volatile, complex, and less controllable environment in and outside USDA. Although the agencies are embedded in a traditional structure, they are subject to new changes, pressures, and interests. In agriculture, the definition and perspective of the public good is changing dramatically. Paul Kennedy (1993) noted that "the task of reconciling technological change and economic integration with traditional political structures, social needs, institutional arrangements, and habitual ways of doing things looms as our greatest problem in the future." That transformation will be met only when USDA redefines itself, how it achieves its mission, and the new relationships among food, health, environment, and society. The future of agriculture will have little resemblance to its past; thus, USDA and its agencies need to create a new 21st-century agenda and an organization to match.

Within USDA, the missions of research and education are still appropriate, but their context is dramatically different. The fundamental concepts of how agencies work, what they work on, and whom they work with are all being called into question, with agency functions, operations, and organization. Future success will engage new leadership, organization models, and ways of conducting work.

The REE organizations need to manage the intellectual capital of their professional staff so that their joint capabilities are complementary and exceed the sum of the capabilities of their separate parts. A key to accomplishing this will center on the need to harness the intelligence and spirit of people at all organizational levels to share and build knowledge continuously. The REE agencies will need to be both equipped and inclined to lead change and work in different and innovative ways. Adopting new models of organizational collaboration will be required for future success (see Chapter 5).

In conclusion, universities, government agencies, and private businesses all have knowledge of immense value. One of the critical questions that has emerged as we shift into the 21st century is how to harness intelligence and ideas from people in various organizations and at all levels of organization to share and build knowledge continuously.

**FINDING: The current organizational structure of research efforts in the REE agencies limits the combined effectiveness of the agencies. The intramural research efforts of the Agricultural Research Service (ARS), the Economic Research Service (ERS), and the National Agricultural Statistics Service (NASS); the competitive grants programs and congressionally mandated grants of the Cooperative State Research, Education, and Extension Service (CSREES); and the federal formula funds of CSREES make up a diverse and diffuse research agenda. Leadership to provide intellectual guidance and a long-term, coherent vision for REE research, promote intra-agency coordination, broker partnerships outside the REE agencies, and integrate REE's research within the federal research program is lacking. No position in the REE administrative structure has the visibility and prestige of the directors of the National Institutes of Health (NIH) and the National Science Foundation (NSF), and the scientific reputation of the REE agencies suffers from this lack.**

**RECOMMENDATION 9: There is a national need for a high-level leader to represent food and agricultural research and to promote opportunities for the research system. Such a leader should be vested with the authority to develop the food and agricultural research agenda, redirect funds to emerging issues and emergency needs, integrate the efforts of the individual agencies, and facilitate collaboration and coordination with scientists outside USDA and elsewhere in the federally supported research system. The leader should be selected on the basis of outstanding scientific and administrative accomplishments and must command the respect of the agricultural community and the broad scientific community.**

The committee considered a number of alternatives for implementing this recommendation, including establishing new positions and strengthening existing positions. The committee discusses below the advantages and disadvantages of four of these alternatives.

1. A new position of research director, reporting directly to the secretary of agriculture, could be established at USDA. The research director would be visible and prestigious, would provide vision and leadership, and would command respect in Congress, in the department, in the administration, and within the public at large. Such a director, when asked, could provide testimony to Congress. A position reporting directly to the secretary would attract a high-stature scientist. A research director could set the strategy for the REE research agenda, broker partnerships outside REE, and galvanize inter–mission-area collaboration. A research director could serve as an additional liaison for hearing action-agency research needs and could respond to needs not being met. The director could also provide guidance on research and data-collection activities conducted or administered at USDA, such as the Forest Service and the Natural Resources Inventory, in the Natural Resources and Environment Mission Area, and the producer-assessment-funded research programs, administered through the Market and Regulatory Programs Mission area. A long-term appointment of 6 years that overlapped presidential elections could foster stability and continuity, stimulate longer-term efforts, and help the research director act as a counterforce to political relationships (such as those resulting in funding to facilities and earmarks) that detract from the REE research strategy. Other research positions in the federal government on which the research-director position might be modeled include the position of NIH director and NSF director. Attributes of the NSF and NIH director positions that could be emulated include the stature and scientific credentials of the positions, their role in coordinating efforts within the organization, and their strong influence over the president's budget. A potential disadvantage of establishing a research-director position would be the loss of program control by agency heads and the undersecretary and the need for shifts in line and budgetary authority. It also is important to note that a 6-year term may have the disadvantage of weakening the position in later administrations. Establishing such a position would require congressional action.
2. The committee considered the option of changing and strengthening the role of the undersecretary for REE to achieve the desired functions. This could be achieved by granting the undersecretary more influence over the budget of the REE agencies, which might make the position more attractive to nominees of high scientific stature. Another mechanism for strengthening the position would involve the granting of some discretion-

ary funds to the undersecretary. Such funds could be used for collaborative activities or for research on new and emerging issues. Relying on the undersecretary position to fulfill leadership needs may be disadvantageous because the undersecretary position is not at the same level as that of the directors of NIH, NSF, and so on, and because the undersecretary may not necessarily be selected on the basis of scientific credentials. A further disadvantage is that the short term of the undersecretary position might lead to lack of long-term vision and continuity in research direction and that it is unlikely that the term could be lengthened to 6 years. The committee believes that since the 1994 reorganization, the undersecretary position has not effectively served the function of brokering partnerships among the REE agencies or with other federal agencies beyond USDA. Furthermore, the committee believes that the undersecretary's role should be more appropriately related to integrating research, education, and extension functions within the mission area rather than to providing the long-term intellectual vision for research.
3. The committee considered the option of strengthening the roles of REE administrators in setting research priorities within the agencies but identified no mechanism for solving the problems of limited coordination among agencies, the separation of budgetary lines among agencies, and competition for budgets. Strengthening the roles of administrators would not solve the problem of identifying overall leadership for the research establishment at USDA. It would not address the issue of fostering collaboration between USDA and other federal research agencies.
4. The committee considered an additional option of establishing a Senate-confirmed associate director for agriculture and natural resources in the Office of Science and Technology Policy. The committee acknowledges that the scope of such a position would be broader than that of the four REE agencies, but it would respond to the needs identified to establish partnerships across the federal agricultural research enterprise. Such a position would be able to influence the president's budget and help to manage and oversee the various collaborations recommended in the report. Such a position could also integrate REE research in the context of the federal agricultural research enterprise. A disadvantage of the position is that it would be far removed from the REE agencies.

After careful consideration of the alternatives, most committee members preferred the first option—the creation of a new position of research director reporting directly to the secretary of agriculture—for establishing the high-profile leadership that is needed to implement the new vision for food and agricultural research described in this report. Several committee members concluded that other options also could successfully address the need for enhanced leadership of the nation's food and agricultural research effort.

Whichever option is chosen, the committee recognizes that an outstanding, high-profile leader is not exclusively responsible for the excellence, visibility, and prestige of an agency. Leaders can change and have changed the strategic direction of some agencies for the benefit of the nation and the world. However, it is important to acknowledge that the sum total of the agencies' scientists and operations is what makes them outstanding in the long run.

## PROFESSIONAL SKILLS, EXPERTISE, AND TRAINING

High-quality scientists are the foundation of high-quality research. One of the best measures of the scientific stature of a research unit is the quality of the people employed in it. The committee considered aspects of the quality and potential for advancement of REE scientists, including an analysis of the REE workforce composition, hiring and recruitment policies, training and opportunities for professional development, and education. Staff performance standards and incentives are considered in Chapter 6, in the context of research quality.

### Workforce Composition

In 2001, 4,132 science-related technical staff were employed by the REE agencies. Almost 75% of these, 3,075, were employed by ARS, the large intramural research effort of USDA. ERS employed 319 science-related technical staff (8%), largely economists; and NASS employed 566 (14%), largely statisticians. Only 163 science-related technical staff (4%) were employed by CSREES; its research role is limited to administration of formula funds and extramural grants. Of the ARS science-related technical staff, 1,980 (48% of the REE total) were PhD research scientists who served as the direct leaders of research projects (USDA, 2001).

Although a list of 50 job categories given to the committee by REE describes duties in a number of fields (see Table 7-1), chemistry, entomology, microbiology, general biology, and genetics accounted for half the ARS professional workforce. A substantial number of the 981 chemists, geneticists, and microbiologists who worked for ARS were probably involved in food safety, food technology, and nutrition, but only 62 were employed with job titles in food technology and 50 in dietetics or nutrition. Similarly, small numbers of science-related technical staff had primary job titles clearly related to environmental science; these included 29 ecologists, 35 range conservationists, 4 environmental engineers, and 2 wildlife biologists. Other science-related technical staff—such as microbiologists, physicists, and hydrologists—were likely to have environmental training but were not immediately identified from workforce titles. Nevertheless, relatively few scientists in the REE agencies appeared to have the broad training required to integrate across levels of ecologic organization and across complex agricultural landscapes—since these are rapidly advancing scientific fields and the agencies have

**TABLE 7-1** REE Professional Employment in Science-Related Occupations, as of June 10, 2001

| Occupation | ARS | ERS | NASS | CSREES | Total REE |
|---|---|---|---|---|---|
| Agricultural engineering | 124 | 0 | 0 | 5 | 129 |
| Agronomy | 99 | 0 | 0 | 0 | 99 |
| Animal science | 68 | 0 | 0 | 4 | 72 |
| Biomedical engineering | 2 | 0 | 0 | 0 | 2 |
| Botany | 20 | 0 | 0 | 0 | 20 |
| Chemical engineering | 31 | 0 | 0 | 1 | 32 |
| Chemistry | 436 | 0 | 0 | 1 | 437 |
| Civil engineering | 26 | 0 | 0 | 0 | 26 |
| Dietetics and nutrition | 50 | 1 | 0 | 4 | 55 |
| Ecology | 29 | 0 | 0 | 1 | 30 |
| Economics | 5 | 297 | 0 | 5 | 307 |
| Electrical engineering | 4 | 0 | 0 | 0 | 4 |
| Electronic engineering | 8 | 0 | 0 | 0 | 8 |
| Entomology | 309 | 0 | 0 | 11 | 320 |
| Environmental engineering | 4 | 0 | 0 | 0 | 4 |
| Fishery biology | 6 | 0 | 0 | 0 | 6 |
| Food technology | 62 | 0 | 0 | 1 | 63 |
| Forestry | 1 | 0 | 0 | 1 | 2 |
| General biological sciences | 278 | 0 | 0 | 84 | 362 |
| General physical sciences | 26 | 0 | 0 | 0 | 26 |
| General engineering | 29 | 0 | 0 | 0 | 29 |
| Genetics | 245 | 0 | 0 | 5 | 250 |
| Geography | 1 | 1 | 1 | 0 | 3 |
| Geology | 5 | 0 | 0 | 0 | 5 |
| Home economics | 6 | 0 | 0 | 0 | 6 |
| Horticulture | 64 | 0 | 0 | 0 | 64 |
| Hydrology | 36 | 0 | 0 | 0 | 36 |
| Industrial hygiene | 2 | 0 | 0 | 0 | 2 |
| Mathematical statistics | 4 | 0 | 90 | 0 | 94 |
| Mathematics | 9 | 0 | 0 | 0 | 9 |
| Mechanical engineering | 14 | 0 | 0 | 0 | 14 |
| Medicine | 1 | 0 | 0 | 0 | 1 |
| Meteorology | 5 | 0 | 0 | 0 | 5 |
| Microbiology | 300 | 0 | 0 | 7 | 307 |
| Materials engineering | 5 | 0 | 0 | 0 | 5 |
| Pharmacology | 5 | 0 | 0 | 0 | 5 |
| Physics | 7 | 0 | 0 | 0 | 7 |
| Physiology | 79 | 0 | 0 | 0 | 79 |
| Plant pathology | 166 | 0 | 0 | 4 | 170 |
| Plant physiology | 205 | 0 | 0 | 4 | 209 |
| Psychology | 2 | 0 | 0 | 1 | 3 |
| Range conservation | 35 | 0 | 0 | 0 | 35 |
| Social science | 0 | 12 | 0 | 17 | 29 |
| Sociology | 0 | 4 | 0 | 4 | 8 |
| Soil science | 182 | 0 | 0 | 1 | 183 |
| Statistics | 14 | 4 | 475 | 0 | 493 |
| Textile technology | 8 | 0 | 0 | 0 | 8 |
| Veterinary medical science | 44 | 0 | 0 | 1 | 45 |
| Wildlife biology | 2 | 0 | 0 | 0 | 2 |
| Zoology | 12 | 0 | 0 | 1 | 13 |
| Total | 3,075 | 319 | 566 | 163 | 4,123 |

Source: REE Office of Human Resources, 2001.

identified so few staff as primarily environmental scientists. Social-science staff other than economists were scarce in REE; there were only 29 noneconomist social-science staff in all of REE. There were proportionally few social scientists of any type in ARS, NASS, and CSREES.

A time series of ARS research scientists from FY 1986 to FY 2001 shows increases in the percentage of research scientists in specific fields, such as microbiology (an increase from 5% in 1986 to 10% in 2001) and genetics (an increase from 6% in 1986 to 10% in 2001); one-fifth of the new research scientists hired in FY 2001 were in molecular biology and genetics. The increase in ecology-research scientists was very small (0.27% in 1986 to 0.9% in 2001). With respect to sex and ethnicity, REE technical staff are predominantly white men. For example, 83% and 88% of ARS scientists are male and white, respectively (see Table 7-2). There is a similar imbalance in sex and ethnic diversity among REE agencies, with some agencies performing better than others in meeting diversity goals.

ARS expects a slight increase in retirement rates in the next 5 years as the research workforce ages. This would provide an opportunity to alter the expertise and diversity of the research staff.

> **FINDING: Staffing is increasing in disciplines suited to exploiting some of the research frontiers—such as molecular biology and genetics—although the increases in ecology are still very small. There is a continuing lack of scientific expertise in the nutritional, environmental, and social sciences and imbalances in ethnicity and sex within and across agencies.**
>
> **RECOMMENDATION 10: REE should increase the hiring of scientists in research fields that have the greatest opportunities to address societal goals. Those include integrative environmental science, ecology, economics, and sociology; human genetics (including statistical human genetics) and bioinformatics; and human nutrition, public health, and**

TABLE 7-2 Demographic Composition of REE Technical Staff

| Agency | Race, % | | | | Sex, % | |
|---|---|---|---|---|---|---|
| | Asian | Black | Hispanic | White | Male | Female |
| ARS | 7.7 | 1.4 | 2.3 | 88 | 83 | 17 |
| CSREES | 6.5 | 8.1 | 0.8 | 84 | 66 | 34 |
| ERS | 8.2 | 3.8 | 1.6 | 86 | 73 | 27 |
| NASS | 3.4 | 12 | 2.9 | 81 | 68 | 32 |

Source: REE, 2001.

food safety. **REE agencies should continue to develop new methods for recruiting and retaining women and members of ethnic minorities.**

Greater balance of scientific disciplines within and between agencies could be achieved by promoting greater interagency cooperation (see Chapter 5).

### Staff Recruitment

The public and private sectors of the economy compete for creative science and technology personnel, and the growth of PhD agricultural-scientist employment has been faster in the private sector than in the public sector since 1973 (NRC, 1988, 1995). Of agricultural-scientist doctoral graduates surveyed in 1996 who planned employment after completing their doctorate (59.3%), 22.8% planned employment in academe, 16.6% planned employment in industry and business, and 12.6% planned employment in government positions (NRC, 1998). REE administrators, in interviews with the committee, indicated that they found it difficult to compete with consulting firms, universities, and other federal agencies for high-quality candidates, although ERS noted that its salaries are only slightly below those of good universities (a $10,000–15,000 difference). Universities may offer greater intellectual freedom and prestige to prospective employees. Our interviews with REE human-resources personnel and administrators indicated that recruiting for diversity is particularly challenging for the REE agencies, given the salary disadvantages. Recruitment and retention procedures are changing for scientists at the most senior level, however. ARS reported to the committee that other federal agencies (NIH and the Food and Drug Administration [FDA]) can offer a $40,000–50,000 advantage in salary to senior scientists who have outstanding reputations in their fields (through the Senior Biomedical Research Service). A similar system, the Senior Scientific Research Service (SSRS), was recently authorized in the 2002 farm bill (US Congress, 2002) to attract and retain scientists at the Nobel Prize level of accomplishment. Regulations and implementation plans are being developed for SSRS. The system will apply to ARS and Forest Service research scientists whose accomplishments, stature, recognition, and impact on scientific theory and knowledge put them above the Government Service-15 pay scale. SSRS is an encouraging step forward and may help to eliminate some of the salary disadvantages at the senior-scientist level.

As conveyed to the committee in its interviews with REE administrators, a second factor with an adverse effect on recruitment is the increasingly large number of non-US citizens receiving PhD degrees in the agricultural sciences; these people are not eligible for employment in US government agencies (Ballenger and Klotz-Ingram, 2000). For example, a survey of 1862 land-grant university colleges of agriculture, renewable natural resources, and forestry indicated that

40% of the doctoral degrees conferred in 1998–1999 were to non-US citizens (FAEIS, 1998/1999). Thus, the pool of applicants from which REE can draw is much smaller than for competing employers.

Hiring procedures are a third factor that influences recruitment. Many young scientists completing PhD degrees or in postdoctoral training are not familiar with the hiring procedures of the REE agencies; these procedures are thus not likely to produce a pool of applicants as large as those of academic institutions, the chief competitors for young research scientists. The procedures for hiring in USDA are established by the Office of Personnel Management (OPM), which oversees employee rules for all federal agencies. OPM has extensive regulations on hiring procedures and qualifications and sets the restrictions and requirements. Each position has published qualification requirements, including basic qualifications (OPM, 2002b) and specialized experience (knowledge, skills, and abilities). Compared with the academic or private-industry job-application systems, the OPM system, which extends to such details as numbers of hours of coursework and specific undergraduate courses, may be more complex and time-consuming for a job applicant. OPM also determines job classifications and categories (OPM, 2002a), and our conversations with REE Office of Human Resources (OHR) staff in REE indicated that the publication of new series[1] definitions occurs very slowly at OPM. For example, the occupation series for ecology was created in 1977, and there is not yet a definition for bioinformatics. The rules can constrain an agency's ability to meet human-resources needs but can be largely circumvented by creative and competitive administrators. The complexity of the requirements may deter many candidates. Rigid and complex hiring requirements are likely to create difficulties in recruiting and hiring professionals who have the multidisciplinary experience needed for meeting contemporary challenges in genomics or environmental research.

The REE agencies have made use of a wide array of recruitment and retention incentives for several years. An interesting innovation in hiring procedures was developed and tested in ARS (and the Forest Service) in 1990–2001 to make recruitment and selection flexible and responsive to local recruitment needs. The Demonstration Project[2] included the development of a different kind of candidate-assessment method and provided the flexibility to use recruitment incentives (see Box 7-1). The project has given managers greater flexibility to adapt to local

---

[1] A series is a subgroup of a group of related occupations that includes all classes of positions at the various skill levels in a particular kind of work. Series are assigned specific numerical codes for purposes of identification and human resources management (OPM, 2001).

[2] According to OPM, "a demonstration project provides a means for testing and introducing beneficial change in Government-wide human resources management systems. A Federal agency obtains the authority from the Office of Personnel Management to waive existing Federal human resources management law and regulations in title 5, United States Code, and title 5, Code of Federal Regulations, to propose, develop, test, and evaluate interventions for its own human resources management system that shape the future of Federal human resource management" (OPM, 2002c).

## BOX 7-1
## The Agricultural Research Service Demonstration Project

ARS and the Forest Service are testing an innovative recruitment system, the Demonstration Project, developed in response to concerns regarding the adequacy of the traditional OPM recruitment and hiring system.

In contrast with the OPM recruitment system, in which a numerical rating and ranking system is used and only three names of qualified job candidates can be released to a supervisor (the so-called "rule of three"), under the Demonstration Project a supervisor can choose from a broader selection of job applicants. The Demonstration Project uses an alternative candidate-assessment method based on eligible and quality groupings of candidates, from whom managers may normally select any candidate referred. ARS supervisors can also offer unlimited monetary recruitment incentives and reimbursement of relocation travel and transportation expenses, subject to higher-level review and approval of fund availability.

Under the Demonstration Project, vacancies can be advertised in the media most appropriate to the geographic location and type of position. Applicants submit applications directly to the ARS location, not to the examining office. ARS has experimented with full-page ads in *Science* magazine three times per year. OHR reported to the committee that its application volume mushroomed after it placed vacancy ads on the *Science* Web site.

In an analysis of the Demonstration Project conducted in 1995 by OHR, several advantages of the project were identified, including improved ability to compete for qualified employees, greater flexibility of hiring procedures, and improved public perception of the agency as an employer and community member. The project was not found to have an adverse effect on groups that historically experienced discrimination. The model was cited in the vice-president's report on the National Performance Review as "a human resource management program consistent with streamlining flexibility, local control, and customer orientation" (Gore, 1993). The Merit Systems Protection Board, an organization that conducts analysis on the civil service system, also conducted a study of the "rule of three" that cited the Demonstration Project authorities. Although there have been efforts to expand the program to other agencies or to the rest of the federal government and to make the project authorities permanent, such changes have not yet been approved by OPM. Thus, Demonstration Project authorities are in 1998 USDA appropriations legislation for ARS and Forest Service use only.

Source: ARS, Office of Human Resources, 2001.

conditions and has improved the public perception of the recruitment process. The REE agencies have also implemented recruitment and retention incentives authorized by the Federal Employees Pay Comparability Act (US Congress, 1990a), including flexible work schedules; flexible workplace programs; transit subsidies; recruitment, retention, and relocation payments; leave-donor programs; family-friendly leave policies; repayment of student loans; and tuition-support programs.

Cooperative-agreement funding mechanisms have also been used at some university-based research centers (Tufts and Baylor Human Nutrition Research Centers, for example) and could be used at all ARS laboratories associated with academic institutions to improve hiring success and to mitigate unnecessary regulatory burdens.

A review of recruitment in ARS conducted in FY 2000 by the REE OHR (USDA, 2001) indicates that the average number of applicants for ARS research-scientist positions was 11, and that 7 of those were referred to center directors for consideration. Although data on the previous number of applicants received are not available from ARS, OHR staff reported an increase in the number of applications for almost every scientist-year position when the consolidated recruitment approach was implemented under the Demonstration Project.

**FINDING: A number of current hiring practices and other government regulations adversely influence the ability of REE to compete for the best scientists. It is encouraging that the REE agencies continue to seek permanent authority to use more-flexible hiring mechanisms, such as the Demonstration Project or cooperative agreements.**

### Postdoctoral Programs and Hiring of New Scientists

A major source of new ARS hires appears to be postdoctoral scientists within the system. These positions provide an opportunity to bring researchers with new expertise into the ARS laboratories and are most common in the larger laboratories or those colocated on college campuses. Support of postdoctoral scientists in the ARS system is available from a systemwide fund or from funds available to research groups. For FY 1998–2000, the average number of postdoctoral scientists in the ARS system was 249, and there were 103 new hires each year. There were also 41 conversions of postdoctoral positions to permanent positions each year, about 25–30% of the new hires each year (see Table 7-3). Table 7-4 shows the funding pattern for the ARS postdoctoral project, which has declined in recent years. Many locally funded postdoctoral scientists are also hired each year by individual management units, and salary is budgeted at the local level. ARS OHR reported an average range of 150 to 200 locally funded positions on the rolls each year, at an average cost of $50,000 per year; the cost (salary and benefits) of locally funded postdoctoral scientists each year ranges from $7.5 million to $10 million.

**TABLE 7-3** ARS Postdoctoral Employment

| Fiscal Year | On Board at End of Year | Number of New Hires | Number of Conversions to Permanent Positions |
|---|---|---|---|
| 1998 | 245 | 105 | 48 |
| 1999 | 262 | 106 | 37 |
| 2000 | 240 | 97 | 38 |
| 2001 | 262 | 67 | 32 |

Source: REE Office of Human Resources, 2001.

**TABLE 7-4** Funding Levels for the ARS Postdoctoral Program, 1985–2002

| Year | Funding, nominal dollars | Funding, constant 2000 dollars[a] |
|---|---|---|
| 1985 | 1,269,000 | 2,199,307 |
| 1986 | 2,160,000 | 3,600,000 |
| 1987 | 3,362,707 | 5,388,954 |
| 1988 | 4,541,885 | 6,892,086 |
| 1989 | 4,496,466 | 6,441,928 |
| 1990 | 4,434,491 | 6,041,541 |
| 1991 | 4,434,491 | 5,873,498 |
| 1992 | 4,434,491 | 5,729,317 |
| 1993 | 4,434,491 | 5,599,105 |
| 1994 | 4,434,491 | 5,401,329 |
| 1995 | 4,434,491 | 5,247,918 |
| 1996 | 4,394,185 | 5,062,425 |
| 1997 | 4,359,804 | 4,860,428 |
| 1998 | 4,333,497 | 4,644,691 |
| 1999 | 4,333,497 | 4,472,133 |
| 2000 | 4,323,497 | 4,323,497 |
| 2001 | 4,323,965 | 4,118,062 |
| 2002 | 4,323,965 | 4,011,099 |

[a]Constant-dollar conversions based on R&D deflator in Table F-11.

Source: ARS, 2002.

The committee believes that a strong postdoctoral program is essential for ensuring the continuous flow of new knowledge, skills, interests, and perspectives into the work of the agencies; for creating important links of new talent from the land-grant institutions with the other REE agencies; for stimulating problem-oriented research on critical emerging subjects; and for providing a pool of young scientists from which the agencies can identify and recruit the next generation of career scientists. The committee is encouraged by the continuing use of postdoctoral fellowships or similar postdoctoral appointments in the REE agencies.

### Training and Opportunities for Professional Development

Opportunities for short-term and long-term training exist in REE. The REE OHR reported to the committee that it offers employees more than 50 training and development programs and courses each year through the Employee Development and Training Program. The courses include leadership development, management skills, computer and presentation skills, congressional and policy issues, new-employee orientation, and related management topics (USDA, 2001). Other short-term training opportunities in substantive or experiential subjects were reported in our interviews with REE administrators. ERS, for example, reported subsidizing inhouse training for econometrics, computer, and statistical software packages; participation in professional meetings; and opportunities for travel detail with the Office of Management and Budget, the US Trade Representative, and the president's Council of Economic Advisors. ARS reported that colocation of laboratories on university campuses provided the opportunity for researchers to serve as adjunct faculty.

Long-term training policies also permit staff to develop comprehensive plans that involve substantial professional training. Although data on continuing education were not available through the REE OHR, the committee learned anecdotally from its interviews with staff that REE employees are taking advantage of continuing university training and the pursuit of PhD degrees with support from their agencies. The committee found little evidence of training or capacity-building in REE on critical current factors affecting the agricultural sciences, such as environmental and natural-resources sciences or consumer and health sciences. To conclude, although the training offerings may be substantial, there are no data relating them to essential core competences or success skills.

**FINDING: REE has made substantial efforts to build internal capacity by promoting training and professional development through cooperation with institutes of higher education, not-for-profit organizations, and the private sector, although there is little evidence of a strong connection between training and mission goals or core competences. Incentives are a promising mechanism for USDA research scientists to further their training in new subjects in the form of sabbaticals, short-term visits, or collaborative projects with people at land-grant universities or private academic and research institutions.**

### Broadening the Scientific Base

Interaction with scientists outside agriculture, such as those in ecology and conservation science, can contribute to meeting the research frontiers identified in Chapter 3 by identifying possible integration with other disciplines, bringing cutting-edge concepts from their fields, and identifying factors that will shape future directions for agricultural research and policy. Scientists outside agriculture would also benefit from such collaboration in that they often lack an awareness of the critical importance of the agricultural research system for supplying technology to improve the compatibility of conservation and agriculture goals. Such broadening of the scientific base was a specific goal of National Research Initiative legislation, in which Congress called for the "widest participation of qualified scientists" in the competitive-grant process (US Congress, 1990b). Although the success of the directive has been mixed, the intent is still seen as a strength of the program (NRC, 2000).

There are a variety of means for encouraging cross-disciplinary exchange. Encouraging and developing mechanisms for scientists to participate in sabbaticals or exchange programs between agencies and organizations would benefit research within and outside REE. Research sabbaticals for scientists from other federal agencies to work in the REE agencies and research sabbaticals for REE scientists in federal agencies or nongovernment organizations would contribute to the exchange. Such exchange would help to create a system in which the benefits of agricultural research are better understood by the broader scientific community and in which cutting-edge thinking in other fields of science is better understood by agricultural researchers. Such mechanisms are already being used by REE agencies; for example, ERS scientists have been detailed to organizations such as the Organization for Economic Cooperation and Development and the Food and Agriculture Organization.

The committee's interviews with REE administrators revealed that agencies are already involving outside expertise in a variety of ways. ARS reported that colocation of ARS laboratories at land-grant universities was an effective mechanism for involving outside expertise in ARS laboratories, CSREES reported that flexible hiring arrangements through Intergovernmental Personnel Act agreements had permitted it to engage 5% of its staff through short- to medium-term assignment, and NASS reported that it offers year-long fellowships to scientists and engages university scientists collaboratively through cooperative agreements.

### REE's Role in Education: Future Human Capacity

REE contributes to building high-quality educational capacity through its support of the research establishment in the land-grant university system. The committee did not undertake a comprehensive review of the capacity within the land-grant system, over which REE has limited influence through administration of formula funds. However, the committee found important evidence that REE is playing a catalytic role in investing in specific programs to develop new expertise, to enhance institutional capacity, and to broaden participation in public agricultural research.

Table 7-5 shows a variety of programs administered by CSREES in expertise development and institutional enhancement. ARS has invested in training and

**TABLE 7-5** Summary of CSREES-Administered Higher-Education Programs

| Program | Purpose | Funding |
| --- | --- | --- |
| 1890 Institution Teaching and Research Capacity Building Grants Program | Competitive program for attracting more students from underrepresented groups into food and agricultural sciences, expanding links among 1890s universities with other colleges and universities and strengthening the teaching and research capacity of the 1890 institutions | FY 2002, $9.479 million |
| Food and Agricultural Sciences National Needs Graduate Fellowships Grants Program | Competitive-grant program for recruiting predoctoral students | FY 2001–2002, $5.6 million |
| Multicultural Scholars Program | Baccalaureate scholarship program for underrepresented racial and ethnic groups | FY 2000–2001, $1.9978 million |
| National Awards Program for Excellence in College and University Teaching in the Food and Agricultural Sciences | Honors outstanding teachers and strengthens instructional programs | FY 2002, Two $5,000 national and eight $2,000 regional awards. |
| Higher Education Challenge Grants Program | Competitive-grant program in undergraduate teaching | FY 2001, $4.350 million; FY2002 $4.058 million |
| Hispanic-Serving Institutions Education Grants Programs | Competitive-grant program to strengthen Hispanic-serving institutions | FY 2002, $3,340,000 |
| Tribal Colleges Endowment Fund | Distributes the interest earned by an endowment to enhance education in agricultural sciences and related fields for Native Americans | FY 2002, $7.1 million |
| Tribal Colleges Education Equity Grants Program | Formula-grant program designed to strengthen higher-education instruction in the food and agricultural sciences in the 1994 land-grant institutions | $50,000/institution upon receipt of a plan of work; FY 2001, $1.5486 million |
| Tribal Colleges Research Grants Program | Competitive-grant program for research that addresses high-priority tribal, national, or multistate areas | FY 2002, $925,000 |

*continued*

**TABLE 7-5** Continued

| Program | Purpose | Funding |
|---|---|---|
| Alaska Native-Serving and Native Hawaiian-Serving Institutions Education Grants Program | Noncompetitive-grant program to strengthen higher education in public or private nonprofit Alaska native-serving institutions and native Hawaiian-serving institutions | FY 2002, $2.997 million |
| Secondary and Two-Year Postsecondary Agriculture Education Challenge Grants Program | Program for public secondary schools and public or private nonprofit community or junior colleges | FY 2002, $798,000 |

Source: USDA, 2002a.

capacity-building by training postdoctoral fellows and graduate students at universities colocated with its facilities and by administering a summer program for graduate students to study in Japan, Korea, and Taiwan, in collaboration with NIH and NSF (USDA, 2002b). ARS also supports a scholarship program for students of 1890 universities in its laboratories; at the end of the scholarship period, the students are able to apply for positions on a noncompetitive basis. ERS conducts a summer-internship program that has strong participation by students in the 1890 institutions. NASS reported to the committee that its summer internship program is a valuable source of new hires. Box 7-2 demonstrates that some of these investments have tried to encourage multidisciplinary training and training in new and emerging fields. Several of the projects are collaborative. There is a clear effort to target underrepresented groups.

## INFORMATION CAPACITY: REE EFFORTS IN DATA MANAGEMENT, COLLECTION, AND SHARING

Information capacity is a critical function in the process of advancing knowledge related to agricultural resources and the application of that knowledge to societal progress. In some cases, the private benefits from a new dataset are large enough for a private firm or individual to undertake the design, collection, and distribution of the data. In other situations, private action will lead to substantial underprovision of datasets and may be a serious constraint in accomplishing research goals. Government agencies, including USDA, are well positioned to collect some types of large datasets that are mandated by legislation or regulations. Furthermore, some types of data—especially those related to food availability, food safety, health, and the environment—are needed for policy decisions

> **BOX 7-2**
> **CSREES Investments in Higher Education**
>
> **The National Needs Fellowship Grant Program**
>
> National Needs Fellowship grants administered in FY 2001–2002 totaled about $5.6 million. National-need fields supported by the program included
>
> - Plant and animal biotechnology.
> - Food, forest-product, and agricultural engineering.
> - Human nutrition and food science.
> - Food, forest-product, and agribusiness marketing or management.
> - Water science.
>
> **The Higher-Education Challenge-Grant Program**
>
> Support was awarded in FY 2001 to a number of universities for development of
>
> - An e-commerce minor.
> - An interdisciplinary and experiential program in tropical agriculture and sustainable development (collaboration among universities).
> - An online farm-animal anatomy course.
> - A food-biotechnology instructional model.
> - A three-dimensional animation of signal-transduction pathways.
> - International nutrition.
> - A Web-based modeling system for soil and water quality.
> - Improved curriculum design for medical nutrition therapy using advice from an expert panel.
> - Web-based instructional materials on exotic-species biology.
> - Curriculum on foreign animal diseases.
> - Faculty training in computer-aided instructional materials at minority-group institutions.
> - A national conference on student critical thinking and writing for faculty and teaching professionals.

and must therefore be in the public domain. Determination of such data needs and funding of data collection and analysis are appropriate and necessary government functions.

Data collection is expensive and USDA must set priorities for data collection. Priorities should be set with consideration given to a number of important criteria:

- That the data help to resolve questions that are considered important by the general population, as opposed to a single business or narrow segments of society.
- That the data can be analyzed and made publicly available quickly enough for the results to be reliable and useful for policy decisions. In some cases, data collection by nongovernment organizations and the private sector might be more cost-effective and comparably useful, provided that the quality of research design and analysis is ensured via USDA contract mechanisms.
- That the data fulfill the needs of agencies that rely on REE for program implementation, evaluation, and policy decisions.
- That the data capitalize on new and emerging electronic technology.
- That the data provide actionable information.

A broader perspective in surveying and collecting data will be necessary to support an expanding and broadening food and agricultural research agenda. Historically, USDA has focused its data-collection responsibilities primarily on production agriculture and secondarily on diet and human nutrition. Today, however, new categories of data and systems for data use are needed. New forms of vertical contracting, concentration in livestock and crop production and marketing, and the effects of global interdependence on food markets, food availability, economic development, public health, disease transmission, climate, and natural resources are changes that will require new datasets.

The mandate to protect water availability, air and water quality, and other environmental resources will call for new techniques for providing data. Spatially integrated data are vital for developing policies that address environmental issues and for measuring policy effectiveness. The committee learned during its interviews with NASS that spatial data on agricultural practices are collected over large geographic areas, but that the data are highly aggregated and not statistically reliable at the local level. Furthermore, compatible environmental-impact data are not collected by NASS or by any other agency. Agroenvironmental indicators show characteristics of the environment over time and provide a means of measuring changes in environmental quality. ERS publishes a set of agroenvironmental indicators—agricultural resources and environmental indicators—but further development of environmental indicators is badly needed.

Alarming increases in the United States and around the world in such diseases as obesity and diabetes require new data, as does the impact of genomics on disease prevention and therapies. The efficacy and safety of newly discovered food ingredients and dietary supplements and the health implications of related changes in food-consumption patterns will demand information that is not now available. The generation of food-composition data and the assessment of the dietary status of the US population are continuing REE responsibilities that attempt to keep pace with the changing food supply and changes in consumer eating behaviors.

The USDA national food-consumption surveys and the Centers for Disease Control and Prevention (CDC) National Health and Nutrition Examination Surveys (NHANES) have served as the cornerstone of the US National Nutrition Monitoring System (LSRO, 1995). REE is also responsible for maintaining the ARS National Nutrient Databank and derivative Standard Reference database (NDSR, 2001) and the Diet and Health Knowledge Survey, which provide information on perceptions of the adequacy of food and nutrition intake, the importance placed on dietary-guidance messages, self-appraised weight status, the importance of factors related to buying food, and beliefs that may influence dietary behavior. The nutrition-monitoring data generated by the national surveys serve as the basis of dietary-guidance and food-assistance programs. The data are needed for agriculture, food, and nutrition policies; food-safety evaluations; exposure assessments by FDA and EPA; food-additive petitions; and applications for pesticides.

A major effort over the last 3 years has been the integration of the dietary portion of the USDA Continuing Survey of the Food Intakes of Individuals with the health portion of the NHANES conducted by the CDC National Center for Health Statistics (NCHS). The merger will result in one national food and nutrition survey that captures the food-consumption expertise of ARS and the health-assessment expertise of NCHS while meeting congressional requests for more-efficient use of government resources.

The data-collection challenge is heightened by the increased need to use information for broader purposes, the increased variety of the users of information, the need to balance the demand for higher-resolution data collection with maintenance of data-provider confidentiality, and the need to balance the increasing demands for data with minimizing the burden on data providers and maintaining survey response rates. New opportunities and challenges in data collection and dissemination are created by electronic communication (existing mechanisms for data collection and dissemination are discussed further in Appendix G). Larger and emerging questions—which require the expertise of social, biologic, and physical sciences—must be answered. Therefore, new and creative approaches to data collection and analysis that integrate the unique strengths and complementary expertise of all the REE agencies, land-grant universities, other government agencies, the private sector, nongovernment and voluntary groups, and international organizations must be implemented. Finally, new technologic tools, including geospatial referencing, are enabling the combination of new and existing datasets from different sources to create new knowledge.

**RECOMMENDATION 11: REE should undertake an analysis of the data development, management, and dissemination needed to support environmental and nutrition policy analysis. REE should work with**

other USDA mission areas to conduct an inventory of available social, economic, biologic, chemical, and physical datasets and to take stock of the data needs of the future. REE should take the initiative in coordinating with other USDA agencies and with other federal agencies to identify where and how data can be more efficiently and effectively used and shared. REE should put into place structures and systems to support data management and dissemination across its agencies.

## INFRASTRUCTURE CAPACITY: RESEARCH FACILITIES

State-of-the-art facilities and equipment are critical requirements for USDA to be able to conduct world-class science and research. Modern facilities are also critical for the recruitment and retention of outstanding scientists, the most important determinant of the future success of REE. Furthermore, scientific progress depends on use of the latest communication and information-technology equipment in sharing knowledge and research findings with other USDA facilities and with private-sector and university laboratories and scientists.

USDA has a substantial infrastructure of research laboratories across the United States, so the cost of operating and maintaining facilities is substantial. The Forest Service and ARS operate most of the facilities. ARS itself has 244 laboratories at 103 locations and 41 worksites. The laboratories include over 3,000 buildings, nearly 70% of which are over 30 years old. The agency also owns 400,478 acres of land dedicated to research (GAO, 2000).

Because of their strong links to the local communities and supportive relationships with legislators at the state and federal level, USDA research facilities seldom close. That social and political reality has a cumulative effect of creating an infrastructure that may be both too large and too expensive. However, long-standing traditions make the USDA facilities system difficult to change.

**FINDING: Maintaining a physical infrastructure that is too large and too expensive will have a major adverse effect on department research unless REE budgets grow substantially or REE is able to gain in efficiency by being permitted to close and consolidate a number of facilities.**

ARS receives a line item annually for repair and maintenance of facilities (in FY 2002, budget authority for ARS facilities was $192 million; Appendix Table F-8a). Laboratory directors and research leaders are required to set aside up to 4% of their program funds to be used for local repair and maintenance projects. Funds specifically for new construction or major renovations and for capital improvements are appropriated by Congress as a separate and independent budget for buildings and facilities.

**FINDING: Maintenance of some facilities has been deferred for many years, and the cost to repair these facilities is mounting to tremendous sums of public funds.**

The 1999 estimate for deferred maintenance of ARS and Forest Service facilities is almost $145 million. Over the next decade, ARS estimates that the cost of repair and maintenance of its facilities will be $874 million (USDA, 1999). The situation is worsened by the likelihood that, because of the long history of inadequate maintenance, it may be unreasonably expensive to repair some facilities to meet modern human-health, employee-safety and environmental building-code requirements.

**FINDING: Current and projected maintenance costs will compete with programmatic research funds, which have increased little over the last decade.**

Because many facilities need major repairs, renovation, and modernization, substantial funds are needed to support these facilities, perhaps to the detriment of the overall USDA research agenda. During difficult budget times and with little actual growth in agriculture-research funding, the costs of facility repair and maintenance for the large infrastructure will continue to be a drain on the budget. These funds may be better directed toward research programs and USDA scientists.

USDA research facilities must accommodate and support agency missions, programs, and goals—not vice versa. They must be considered a means rather than ends in themselves. Furthermore, USDA cannot afford to allow physical facilities and geographic locations to determine the direction of research; research must be driven by the needs of society and by scientific judgments regarding opportunities for critical advancement of knowledge.

**FINDING: Congressional and stakeholder pressures greatly hinder ARS's ability to close some facilities that do not cost-effectively contribute to USDA's national research agenda.**

In 1999, a Strategic Planning Task Force on USDA Research Facilities was directed by Congress to review the department's research facilities, issue a 10-year strategic plan for USDA on facilities, and make recommendations to ensure that a comprehensive research capacity is maintained (USDA, 1999). The task force concurred with previous General Accounting Office reports and recommended that closing, renovating, and consolidating some of the federal laboratories could add greatly to the efficiency and effectiveness of the agency's research. The committee observes that Congress has also highlighted the importance of security upgrades for agricultural research facilities in the Farm Security and Rural Investment Act of 2002 (US Congress, 2002).

**RECOMMENDATION 12: The committee recommends that REE use objective criteria to decide which USDA facilities merit investment of**

budget resources for repair, modernization, or security improvement and which should be consolidated or closed because they are incapable of cost-effectively contributing to the REE research strategy without renovation. These criteria should be established in the public interest and mutually agreed on by key members of Congress and state and local legislators, as articulated in the principles and recommendations of the 1999 Report of the Strategic Planning Task Force on USDA Research Facilities. The closing, consolidation, or renovation of facilities should be implemented.

## SUMMARY

This chapter has considered four dimensions of the REE mission area: organizational capacity, human capacity, information capacity, and infrastructure capacity. A need was identified for leadership to provide intellectual guidance and a long-term vision for REE research, and several options for meeting this need were considered. In describing REE human capacity, the chapter analyzed the REE workforce, hiring and recruitment policies, training and opportunities for professional development, and education. Staff hiring in key research fields was identified as a way to meet research challenges of the future. REE's efforts in data management, collection, and sharing were presented and discussed. A broader perspective in surveying and collecting data will be necessary to support a broadening food and agricultural research agenda, and an inventory of existing data and an analysis of data needs to support future research were recommended. Finally, the status and cost of maintaining the physical infrastructure of REE were discussed, and it was recommended that criteria to determine which facilities should be repaired, consolidated, or closed be developed and used.

## REFERENCES

Ballenger, N., and C. Klotz-Ingram. 2000. Assessing US benefits of training foreign agricultural scientists. Pp 304–321 in Public-Private Collaboration in Agricultural Research, K.O. Fuglie and D.E. Schimmelpfennig, eds. Ames, IA: Iowa State University Press.
FAEIS (Food and Agricultural Education Information System). 1998/1999. Available online at http://faeis.tamu.edu/.
GAO (US Government Accounting Office). 2000. Agricultural research: USDA's response to recommendations to strengthen the Agricultural Research Service's programs and facilities. GAO/RCED-OO-85R ARS Programs and Facilities. Letter to John Kasich, Chairman, Committee on the Budget, House of Representatives. February 15.
Gore, A. 1993. From Red Tape to Results: Creating a Government that Works Better and Costs Less. Washington, DC: US Government Printing Office.
Kennedy, P. 1993. Preparing for the Twenty-First Century. New York: Random House.

LSRO (Life Sciences Research Office). 1995. (The Second Report on) Nutrition Monitoring in the United States: An Update Report on Nutrition Monitoring. Bethesda, MD: Life Sciences Research Office.

NDSR (Nutrient Database for Standard Reference). 2001. Release 13. USDA/ARS. Washington, DC: Agricultural Research Service, US Department of Agriculture. Available online at http://www.nal.usda.gov/fnic/foodcomp.

NRC (National Research Council). 1988. Committee on Evaluation of Trends in Competency Needs in Agricultural Research at the Doctoral and Postdoctoral Personnel Level.

NRC (National Research Council). 1995. Colleges of Agriculture at the Land Grant Universities: A Profile. Washington, DC: National Academy Press.

NRC (National Research Council). 1998. 1996 Doctoral Recipients from United States Universities. Washington, DC: National Academy Press.

NRC (National Research Council). 2000. National Research Initiative: A Vital Competitive Grants Program in Food, Fiber, and Natural-Resources Research. Washington, DC: National Academy Press.

OPM (US Office of Personnel Management). 2001. Handbook of Occupational Groups and Families. August. Washington, DC: Office of Personnel Management. Available online at http://www.opm.gov/fedclass/gshbkocc.pdf.

OPM (US Office of Personnel Management). 2002a. General Schedule Position Classification. Washington, DC: Office of Personnel Management. Available online at http://www.opm.gov/fedclass/html/gsseries.htm.

OPM (US Office of Personnel Management). 2002b. Qualification Standards for General Schedule Positions. Operating Manual HX118. Washington, DC: Office of Personnel Management. Available online at http://apps.opm.gov/publications/pages/default_list_man.htm.

OPM (US Office of Personnel Management). 2002c. What is a Demonstration Project? Washington, DC: Office of Personnel Management. Available online at http://www.opm.gov/demos/index.htm.

US Congress. 1990a. P.L. (Public Law) 101-509. The Federal Employees Pay Comparability Act (FEPCA) of 1990.

US Congress. 1990b. P.L. (Public Law) 101-624. Food, Agriculture, Conservation, and Trade Act of 1990.

US Congress. 2002. P.L. (Public Law) 107-171. Farm Security and Rural Investment Act of 2002.

USDA (US Department of Agriculture). 1999. Report of the Strategic Planning Task Force on USDA Research Facilities. USDA Publication August 1999; prepared by the Strategic Planning Task Force on Research Facilities, Bruce Andrews, Chair. Washington, DC: US Department of Agriculture.

USDA (US Department of Agriculture). 2001. Data submitted to the National Research Council Committee on Opportunities in Agriculture from the Agricultural Research Service. Washington, DC: Agricultural Research Service, US Department of Agriculture.

USDA (US Department of Agriculture). 2002a. Higher Education Programs. Washington, DC: Cooperative State Research, Extension, and Education Service, US Department of Agriculture. Available online at http://www.reeusda.gov/serd/hep/progdes.htm.

USDA (US Department of Agriculture). 2002b. Summer Internships in Japan, Korea, and Taiwan. Washington, DC: Agricultural Research Service, US Department of Agriculture. Available online at http://www.nsf.gov/pubs/1999/nsf99152/nsf99152.pdf.

# 8

# Coda

The future direction of agricultural research will be challenging. The increased economic, social, and ecologic demands on agriculture generate a complex environment for research planning. Although those demands create the opportunity for enhanced social return from agricultural research, they will also tax the ability of the system in many dimensions. There will be trade-offs among research goals that must be addressed with inadequate resources. There will be conflicting signals from traditional and new stakeholders in the agricultural research system. Sometimes research will be called on to resolve trade-offs or perceived trade-offs among the various demands on the agricultural system. Research may also be recruited to mitigate the unforeseen impacts of food and agricultural policies. To meet new demands, established processes and partnerships in agricultural research must evolve without losing their unique value. Those tensions in the research agenda can be managed only through sustained vision, leadership, and political will.

The committee does not underestimate the magnitude of challenges or obstacles in addressing the new demands. In preparing this report, we moved from identifying research frontiers to considering research institutions and processes that will support research at the frontier. As we have shown, much progress has been made in moving forward to address the frontiers and to embrace institutional change, but much remains to be done. Nevertheless, the committee sees many indicators that the vision outlined in this report is feasible.

As this report goes to press, new farm legislation has just been enacted. The research title shows some new initiatives congruent with our vision. Authorized increases in competitive-grants programs—which may not necessarily be realized—signal the perceived value of a flexible, cutting-edge research program that addresses problems of national importance. New mandates and in some cases new funding are identified for biosecurity, biotechnology risk assessment, and organic farming. A new system for recognizing and rewarding scientific excellence has been created. And, in addition to those items in the new legislation, new coalitions of stakeholders are forming to carry their research demands to Congress. Clearly, many of the frontiers identified in this report are receiving increased congressional attention. Yet many of the changes identified by the committee are within the purview of existing budgets and institutional authority and need not wait for congressional action. The vision in this report can be embraced at all levels of the agricultural-research system.

As elements of the premier agricultural-research system on the globe, the US Department of Agriculture (USDA) and its partners have been widely emulated. The increasingly international character of research benefits means that USDA's future choices will have global consequences. Partners in the research effort are increasingly diverse and far-flung, and how USDA chooses to partner with other institutions will provide models for global collaboration. USDA can lead the way for institutional change that responds to new demands on the agricultural system.

# Appendixes

# Appendix A

# S.1150.1998. Agricultural Research, Extension, and Education Reform Act of 1998

The following text is drawn from the 1998 Agricultural Research, Extension, and Education Reform Act of 1998, which mandated the National Academy of Sciences Study. Through subsequent negotiations with USDA, the statement of task for the study panel was broadened to what is currently in the preface and Executive Summary.

**Subtitle C—Studies**

SEC. 632.  STUDY OF FEDERALLY FUNDED AGRICULTURAL RESEARCH, EXTENSION, AND EDUCATION.

(a) Study.—Not later than January 1, 1999, the Secretary of Agriculture shall request the National Academy of Sciences to conduct a study of the role and mission of federally funded agricultural research, extension, and education.

(b) Requirements.—The study shall—
  (1) evaluate the strength of science conducted by the Agricultural Research Service and the relevance of the science to national priorities;
  (2) examine how the work of the Agricultural Research Service relates to the capacity of the agricultural research, extension, and education system of the United States;

(3) examine the appropriateness of the formulas for the allocation of funds under the Smith-Lever Act (7 U.S.C. 341 et seq.) and the Hatch Act of 1887 (7 U.S.C. 361a et seq.) with respect to current conditions of the agricultural economy and other factors of the various regions and States of the United States and develop recommendations to revise the formulas to more accurately reflect the current conditions; and

(4) examine the system of competitive grants for agricultural research, extension, and education.

(c) Reports.—The Secretary shall prepare and submit to the Committee on Agriculture of the House of Representatives and the Committee on Agriculture, Nutrition, and Forestry of the Senate—

(1) not later than 18 months after the commencement of the study, a report that describes the results of the study as it relates to paragraphs (1) and (2) of subsection (b), including any appropriate recommendations; and

(2) not later than 3 years after the commencement of the study, a report that describes the results of the study as it relates to paragraphs (3) and (4) of subsection (b), including the recommendations developed under paragraph (3) of subsection (b) and other appropriate recommendations.

# APPENDIX B

# Subcommittee Statements of Task

The synthesis committee, the Committee on Opportunities in Agriculture, established statements of task for each of the three subcommittees:

1. Identify the priorities for future research and relevance of the agricultural research, knowledge transfer, and capacity-building activities conducted by the US Department of Agriculture Research, Education, and Economics mission area, given modern challenges and the dynamic nature of agriculture.
2. Broadly evaluate the quality, impact, and productivity of current and past research, knowledge transfer, and capacity-building activities.
3. Consider the following questions:
    a. What are important differences between past and future needs?
    b. Does the current research have broad impacts in our society, and is it demand-driven (client-driven) through stakeholder or citizen participation?
    c. How do the quality, impact, and productivity of REE research in a particular field compare with those of research performed through alternative government or private-sector support?
    d. Is there an appropriate balance between basic and applied research, intramural and extramural research mechanisms, competitive and formula- funding mechanisms, and federal and state-run research programs?
    e. How integrative and interdisciplinary is the research? Is the research complementary across the REE agencies?

f. What professional skills, expertise, and training programs are necessary for achieving the research, extension, and education goals needed to achieve REE's desired outcomes?

In answering those questions, the Subcommittee on Economic and Social Development in a Global Context considered the following subjects: the future structure of agriculture; new market opportunities vis-à-vis new information technology; food and agricultural policy; implications of population and income growth; children, youth, the aging, families, and communities; international development, trade markets, and US competitiveness; and new and value-added products.

The Subcommittee on Environmental Quality and Natural Resources considered the following subjects: conservation of soil, water, atmospheric, and biologic resources, including agrobiodiversity; livestock and range management issues; aquaculture; nonnative and invasive species; hydrologic issues, including surface water, subsurface water, and aquifer issues; use of chemicals and biocontrols; waste-management issues; energy resources, including biobased resources; forest resources; land preservation; land-use and land-use change issues; the rural-urban interface and the interface between agriculture and protected areas; climate change; carbon sequestration; land-grant–sea-grant issues; and environmental implications of trade.

The Subcommittee on Food and Health considered the following subjects divided by two major themes. With respect to *production agriculture,* it considered production systems across a wide range of commodities; appropriateness of technologies; implications of functional genomics; implications of precision agriculture, forecasting technologies, and other spatial information tools; consumer-driven preferences; implications of research choices for consumers; and energy sources and costs. With respect to *food safety, diet, and nutrition,* it considered food safety (microorganisms, toxic substances, and food produced from transgenic organisms); health promotion through diet; nutritional enhancement through processing and production; nutrition education; diet-disease links; nutrient-gene interactions (human, plant, and animal genomics and nutrition); allergens; and food additives and interactions.

# APPENDIX C

# A National Research Council Public Workshop

OPPORTUNITIES IN AGRICULTURE: A VISION FOR USDA'S
FOOD AND AGRICULTURAL RESEARCH IN THE 21ST CENTURY

May 22–23, 2001
9:00 am to 4:30 pm
Green Building, Room 104
2001 Wisconsin Avenue, NW
Washington, D.C. 20418

## WORKSHOP AGENDA

**Tuesday, May 22, 2001**

**Session 1:** 9:00–10:45 am
**Title:** Future Views
**Moderator:** Franklin Loew, President, Becker College
**Discussant:** Barbara Glenn, Federation of Animal Science Societies

*Kate Clancy*, Henry A. Wallace Center for Agricultural & Environmental Policy at Winrock International
  *Topic*: Future agriculture and food systems (including organic farming)

*Montague Demment*, Director, Global Livestock Collaborative Research Support Program, University of California, Davis
   *Topic*: Globalization: Revolution and evolution for American agriculture

*Marilyn Jorgensen*, Jorg-Anna Farms Partnership
   *Topic*: Research needs of production agriculture

*Walter Armbruster*, Farm Foundation
   *Topic*: Research needs for agricultural alternatives

*Anne Sydnor*, Food Marketing Institute
   *Topic*: Grocery stores and future food systems, including impact of technology such as shopping online

**Session 2:** 11:00 am–12:45 pm
**Title:** Unifying Research Issues
**Moderator:** William Ogren, Retired Research Leader, Agricultural Research Service, USDA
**Discussant:** Charles Krueger, Department of Agronomy, Pennsylvania State University

*Dick Amerman*, Agricultural Research Service, USDA
*Mike O'Neill*, Cooperative State Research Education and Extension Service, USDA
*Gerald Larson*, Office of Budget and Program Analysis, USDA
   *Topic*: Panel discussion on interdisciplinary research—A success story (Water Quality Project)

*Jerry Gillespie*, Joint Institute for Food Safety Research
   *Topic*: Pulling agencies together for joint research efforts—A success story in the making?

*Fran Pierce*, Washington State University
   *Topic*: Precision agriculture, bioinformatics, forecasting technologies (include relationship between food, feed, fiber, and energy)

*Jill Auburn*, Sustainable Agriculture Research and Education Programs, CSREES, USDA
   *Topic:* Discovering and extending information — Retooling the system

*George Norton*, Virginia Tech
   *Topic*: Impact assessment and tools for evaluating research productivity and quality

**LUNCH BREAK** *(available in basement Refectory)*
12:45–1:45 pm

**Session 3:**   1:45–3:15 pm
**Title:**        Selected Food and Health Topics
**Moderator:**   Susan Harlander, President, BIOrational Consultants, Inc.
**Discussant:**  Donna Porter, Specialist in Life Sciences, Congressional Research Service

*Roger A. Sunde*, Nutritional Sciences, University of Missouri
   *Topic*: Research needs for human nutrition (including effects of genomics)

*Caroline Smith-DeWaal*, Center for Science in the Public Interest
   *Topic*: Consumer concerns about agriculture research

*Catherine E. Woteki*, former Undersecretary for Food Safety, USDA
   *Topic*: Research structure and ethics leading to food systems for healthy populations

*Clare Hasler*, University of Illinois, Urbana-Champaign
   *Topic*: Functional foods and impact on health and society

**Session 4:**   3:30–5:00 pm
**Stakeholder Open Forum** (pre-registered speakers) (10 min. time limit)
**Moderator:**   Susan Harlander, President, BIOrational Consultants, Inc.

*Steve Derrenbacher*, Northeast Pasture Research and Extension Consortium, Woodsboro, MD
*Karl Glasener*, CoFARM, Washington, DC
*Bob Hedberg*, Weed Science Society of America, Washington, DC
*Robert Donaldson*, American Society of Plant Physiologists, Rockville, MD
*Terry Wolf*, President, National Coalition for Food and Agriculture Research, Homer, IL

**Wednesday, May 23, 2001**

**Session 5:**   9:00–10:45 am
**Title:**        Economic and Social Development
**Moderator:**   Carol Keiser, President, C-BAR Cattle Company, Inc.
**Discussant:**  Charles Riemenschneider, Food and Agriculture Organization of the United Nations

***Cornelia Flora***, Iowa State University
    *Topic*: Structure of agriculture—Trends and needs including antitrust, industry consolidation, small farmer survival

***Bruce L. Gardner***, University of Maryland
    *Topic*: Labor migration issues, implications of increase in meat demand, trends toward niche markets, global climate change and shift in production patterns

***Louis Swanson***, Colorado State University (presentation delivered by Cornelia Flora)
    *Topic*: Beyond agriculture: New policies for rural america

***Walter A. Hill***, Tuskegee University
    *Topic*: Agricultural research concerns of underserved populations, particularly in the South

***Neil Cowen***, Dow Agro
    *Topic*: Technological choices for tomorrow, research in industry compared to REE, relationship of agriculture/food to pharmaceuticals

**Session 6:**    11:00 am–12:45 pm
**Stakeholder Open Forum** (pre-registered speakers) (10 min. time limit)
**Moderator:**    Carol Keiser, President, C-BAR Cattle Company, Inc.

*Jere Downing*, Cranberry Institute, Wareham, MA
*Robert Earl*, National Food Processors Association, Washington, DC
*Esther Myers*, American Dietetic Association, Chicago, IL
*Charles Scifres*, Texas Agriculture Experiment Station, Texas A&M University, College Station, TX
*Stephanie A. Smith*, Institute of Food Technologists, Washington, DC
*Tamera Wagester*, Council on Food, Agriculture & Resource Economics, Alexandria, VA

***LUNCH BREAK** (available in basement Refectory)*
12:45–1:45 pm

**Session 7:**    1:45–3:15 pm
**Title:**    Environmental Quality and Natural Resources
**Moderator:**    Phil Robertson, Michigan State University
**Discussant:**    LaReesa Wolfenbarger, University of Nebraska, Omaha

***Ann Sorensen***, American Farmland Trust
   *Topic*: Research needs to support conservation practices for farmers

***Kim Leval***, Consortium for Sustainable Agriculture Research and Education
   *Topic*: Research needs to support sustainable agriculture

***Mike Williams***, Animal and Poultry Waste Management Center, North Carolina State University
   *Topic*: Research needs for problem solving in animal waste handling systems

***Rattan Lal***, Ohio State University
   *Topic*: Soils—Challenges and research needs

**Session 8:**  3:30–5:00 pm
**Stakeholder Open Forum** (pre-registered speakers) (10 min. time limit)
**Moderator:**  Phil Robertson, Michigan State University

*John B. Adams*, National Milk Producers Federation, Washington, DC
*Richard A. Herrett*, Agricultural Research Institute, Washington, DC
*Myron Johnsrud*, Extension and Outreach Programs, National Association of State Universities and Land Grant Colleges, Washington, DC
*Randall E. Torgerson*, USDA, Rural Business-Cooperative Service, Washington, DC

# APPENDIX D

# REE Administrator Interviews

The committee conducted telephone interviews with administrators from the Agricultural Research Service (ARS), the Cooperative State Research, Education, and Extension Service (CSREES), the Economic Research Service (ERS), and the National Agricultural Statistics Service (NASS) in June and July of 2001. The following questions were used to guide the interviews.

**Vision for the Future of Agricultural Research**
- In what directions would you like to lead your agency?
- What factors, both internal and external to REE, help and hinder your moving in those directions?

**Research Priorities**
- What changes, if any, would you like to see in how priorities are established?
- What changes, if any, do you envision for the way in which stakeholders relate to your agency?
- Do you have the flexibility needed to shift resources to newly emerging priorities? If not, what mechanisms would you recommend to improve flexibility?

**Research Quality and Relevance**
- From your perspective, what are some of the most effective ways to ensure research quality and relevance? How are these implemented at your agency?

## Interdisciplinary Research; Interagency Research
- What is your perspective on interdisciplinary research? Interagency research?
- What are some of the barriers to conducting interdisciplinary research within your agency? Between your agency and other federal agencies?

## REE Organization
- Since the reorganization of REE, how do you believe the research agencies are functioning in relationship to USDA action agencies?

## Relationship of Public-Sector Research to Private-Sector Research
- From your perspective, has public research changed by working more closely with industry? If so, how?
- From your perspective, what is the appropriate relationship between your agency and private industry?

## Professional Development and Human Resources
- How can your agency best attract and retain research leadership?
- What changes, if any, in your agency's professional staff are necessary to meet research priorities?
- Do you feel that your agency has sufficient flexibility to make personnel changes?
- In what ways should REE be interacting with universities to ensure that appropriate professionals are available in the future and that current staff have access to scientific innovation?

## Expectations from the NRC Report
- What recommendations and advice from the NRC study panel would be most useful to you in leading your agency?
- What questions have we not asked that you think would be important for producing a forward-thinking, helpful report?

## Questions to the Committee Members from the Administrators

## Closing Remarks

# APPENDIX E

# Action-Agency Administrator Interviews

The committee conducted telephone interviews with administrators from the US Department of Agriculture Animal and Plant Health Inspection Service (APHIS), Farm Services Agency (FSA), Food Safety and Inspection Service (FSIS), Food and Nutrition Service (FNS), and Natural Resources Conservation Service (NRCS) in January and February of 2002. The following questions were used to guide the interviews.

**Welcome and Introductions**

**Background for Telephone Call**
- Questions from Administrators

**Questions from Committee Members to Administrators**
- How do you interact with the REE agencies? What improvements do you suggest?
- How do you make your research needs known to the agencies?
- How responsive and timely are the REE agencies in meeting your needs and requests? What improvements do you suggest?
- How are the quality and usefulness of the responses? What improvements do you suggest?
- What methods do you use to assess the responsiveness, timeliness, quality, and usefulness of REE support to your needs?
- Will the REE agencies have the capacity to meet your future needs?
- Do you believe there are better ways or improved mechanisms to support the research needs of your agency? If so, what?

# Questions to the Committee Members from the Administrators

# Closing Remarks

# APPENDIX F

# Agricultural-Research Funding

**TABLE F-1** Research, Education, and Economics by Agency for FY 1985–2001 Actual and FY 2002 Estimate

Budget Authority, millions of nominal dollars

| Year | ARS | CSREES | ERS | NASS | Total |
|---|---|---|---|---|---|
| 1985 | 528 | 648 | 47 | 58 | 1,281 |
| 1986 | 500 | 614 | 44 | 56 | 1,214 |
| 1987 | 533 | 718 | 45 | 58 | 1,354 |
| 1988 | 572 | 708 | 48 | 61 | 1,389 |
| 1989 | 601 | 703 | 50 | 64 | 1,418 |
| 1990 | 621 | 755 | 51 | 67 | 1,494 |
| 1991 | 689 | 852 | 54 | 76 | 1,671 |
| 1992 | 754 | 927 | 59 | 83 | 1,823 |
| 1993 | 724 | 910 | 59 | 81 | 1,774 |
| 1994 | 757 | 933 | 55 | 82 | 1,827 |
| 1995 | 772 | 931 | 54 | 81 | 1,838 |
| 1996 | 751 | 911 | 53 | 81 | 1,796 |
| 1997 | 800 | 949 | 54 | 100 | 1,903 |
| 1998 | 845 | 859 | 72 | 118 | 1,894 |
| 1999 | 871 | 924 | 63 | 104 | 1,962 |
| 2000 | 903 | 1,091 | 64 | 100 | 2,158 |
| 2001 | 1,019 | 1,150 | 68 | 101 | 2,338 |
| 2002 | 1,247 | 1,033 | 70 | 114 | 2,464 |

Note: ARS estimates include $17 million made available under the Agricultural Risk Protection Act in 2001 and $113 million under Emergency Supplemental to Respond to Terrorism in 2002, consisting of $40 million for research and $73 million for facilities. Constant dollar estimates based on deflators for research expenditures (see Table F-11).

Budget Authority, millions of constant dollars—2000 = 1.00

| Year | ARS | CSREES | ERS | NASS | Total |
|---|---|---|---|---|---|
| 1985 | 915 | 1,123 | 81 | 101 | 2,220 |
| 1986 | 833 | 1,023 | 73 | 93 | 2,022 |
| 1987 | 854 | 1,151 | 72 | 93 | 2,170 |
| 1988 | 868 | 1,074 | 73 | 93 | 2,108 |
| 1989 | 861 | 1,007 | 72 | 92 | 2,032 |
| 1990 | 846 | 1,029 | 69 | 91 | 2,035 |
| 1991 | 913 | 1,128 | 72 | 101 | 2,214 |
| 1992 | 974 | 1,198 | 76 | 107 | 2,355 |
| 1993 | 914 | 1,149 | 74 | 102 | 2,239 |
| 1994 | 922 | 1,136 | 67 | 100 | 2,225 |
| 1995 | 914 | 1,102 | 64 | 96 | 2,176 |
| 1996 | 865 | 1,050 | 61 | 93 | 2,069 |
| 1997 | 892 | 1,058 | 60 | 111 | 2,121 |
| 1998 | 906 | 921 | 77 | 126 | 2,030 |
| 1999 | 899 | 954 | 65 | 107 | 2,025 |
| 2000 | 903 | 1,091 | 64 | 100 | 2,158 |
| 2001 | 970 | 1,095 | 65 | 96 | 2,226 |
| 2002 | 1,157 | 958 | 65 | 106 | 2,286 |

**TABLE F-2** Total R&D by Agency, FY 1976–2003

| | Budget Authority, millions of 2000 dollars, FY | | | | | | | | | | | | |
|---|---|---|---|---|---|---|---|---|---|---|---|---|---|
| | 1976 | 1977 | 1978 | 1979 | 1980 | 1981 | 1982 | 1983 | 1984 | 1985 | 1986 | 1987 | 1988 |
| DOD | 32,685 | 35,317 | 34,223 | 34,100 | 32,991 | 37,180 | 41,948 | 45,967 | 50,976 | 55,421 | 58,334 | 59,512 | 57,854 |
| NASA | 11,458 | 11,677 | 11,570 | 11,997 | 12,336 | 11,827 | 9,233 | 5,311 | 5,464 | 6,172 | 6,214 | 6,943 | 7,112 |
| DOE | 9,964 | 12,782 | 14,479 | 14,321 | 13,574 | 13,182 | 10,877 | 10,029 | 10,286 | 10,361 | 9,234 | 8,690 | 8,834 |
| NIH[a] | 7,348 | 7,640 | 8,136 | 8,479 | 8,107 | 7,644 | 7,437 | 7,870 | 8,435 | 9,125 | 9,021 | 10,261 | 10,498 |
| NSF | 2,172 | 2,228 | 2,261 | 2,205 | 2,146 | 2,049 | 1,960 | 2,056 | 2,277 | 2,460 | 2,333 | 2,447 | 2,408 |
| USDA | 1,682 | 1,671 | 1,814 | 1,884 | 1,682 | 1,749 | 1,670 | 1,718 | 1,782 | 1,737 | 1,600 | 1,750 | 1,695 |
| Interior | 1,055 | 1,006 | 1,040 | 1,079 | 965 | 882 | 775 | 752 | 664 | 674 | 639 | 654 | 646 |
| DOT | 1,045 | 1,018 | 1,069 | 921 | 934 | 872 | 628 | 721 | 892 | 761 | 631 | 495 | 477 |
| EPA | 886 | 898 | 1,086 | 1,011 | 805 | 803 | 573 | 451 | 454 | 529 | 539 | 552 | 567 |
| DOC | 750 | 778 | 805 | 874 | 842 | 739 | 620 | 631 | 662 | 679 | 658 | 669 | 622 |
| Other[a] | 2,561 | 2,741 | 3,073 | 3,263 | 3,368 | 3,003 | 2,223 | 2,243 | 2,064 | 1,742 | 1,832 | 2,010 | 1,901 |
| Total R&D | 71,606 | 77,756 | 79,556 | 80,134 | 77,750 | 79,930 | 77,944 | 77,749 | 83,956 | 89,661 | 91,035 | 93,983 | 92,614 |
| R&D: | | | | | | | | | | | | | |
| Defense | 35,925 | 38,394 | 37,988 | 37,632 | 36,306 | 40,732 | 46,073 | 49,803 | 55,653 | 60,163 | 62,740 | 63,835 | 62,173 |
| Nondefense | 35,682 | 39,363 | 41,568 | 42,503 | 41,445 | 39,197 | 31,871 | 27,948 | 28,302 | 29,497 | 28,295 | 30,147 | 30,441 |
| Basic research | 8,016 | 8,898 | 11,410 | 11,821 | 12,019 | 10,936 | 10,946 | 12,122 | 12,891 | 13,510 | 13,602 | 14,446 | 14,425 |

NOTE: Includes conduct of R&D and R&D facilities. Constant dollar estimates based on deflators for research expenditures (see Table F-11).
[a]Between FY 1991 and 1992, R&D from ADAMHA (HHS) transferred to NIH. ADAMHA R&D included in NIH totals for all years.
[b]Latest estimate for FY 2002.
[c]AAAS estimates of president's FY 2003 request.

Source: AAAS Reports I through XXVI, based on OMB and agency R&D budget data.

| 1989 | 1990 | 1991 | 1992 | 1993 | 1994 | 1995 | 1996 | 1997 | 1998 | 1999 | 2000 | 2001 | 2002[b] | 2003[c] |
|---|---|---|---|---|---|---|---|---|---|---|---|---|---|---|
| 55,270 | 51,593 | 49,283 | 48,810 | 49,051 | 43,252 | 41,833 | 41,224 | 41,514 | 40,267 | 40,132 | 39,960 | 40,705 | 46,047 | 48,866 |
| 8,481 | 9,689 | 10,760 | 11,039 | 11,130 | 11,456 | 11,194 | 10,867 | 10,426 | 10,451 | 10,028 | 9,494 | 9,417 | 9,491 | 9,515 |
| 8,961 | 9,520 | 9,720 | 10,508 | 9,399 | 8,248 | 7,593 | 7,227 | 6,931 | 6,807 | 7,189 | 6,956 | 7,365 | 7,756 | 7,418 |
| 10,817 | 11,045 | 11,891 | 12,434 | 12,488 | 12,757 | 12,736 | 13,162 | 13,620 | 14,051 | 15,475 | 17,234 | 18,864 | 21,146 | 23,576 |
| 2,524 | 2,355 | 2,508 | 2,547 | 2,542 | 2,732 | 2,836 | 2,755 | 2,703 | 2,680 | 2,755 | 2,931 | 3,162 | 3,271 | 3,254 |
| 1,632 | 1,667 | 1,843 | 1,963 | 1,852 | 1,862 | 1,760 | 1,714 | 1,735 | 1,673 | 1,698 | 1,776 | 2,077 | 2,165 | 1,888 |
| 687 | 723 | 821 | 831 | 820 | 862 | 791 | 658 | 659 | 574 | 515 | 618 | 592 | 612 | 560 |
| 467 | 482 | 545 | 802 | 784 | 781 | 788 | 692 | 682 | 632 | 627 | 607 | 684 | 722 | 656 |
| 569 | 578 | 608 | 636 | 627 | 716 | 656 | 556 | 663 | 683 | 690 | 558 | 546 | 549 | 559 |
| 611 | 617 | 711 | 754 | 1,002 | 1,245 | 1,323 | 1,112 | 1,075 | 1,169 | 1,118 | 1,174 | 981 | 1,017 | 980 |
| 2,077 | 2,257 | 2,366 | 2,622 | 2,386 | 2,661 | 2,453 | 2,067 | 2,415 | 2,408 | 2,509 | 2,461 | 2,783 | 2,910 | 2,593 |
| 92,096 | 90,525 | 91,056 | 92,946 | 92,081 | 86,572 | 83,963 | 82,034 | 82,424 | 81,395 | 82,736 | 83,769 | 87,176 | 95,686 | 99,865 |
| 59,569 | 55,829 | 53,477 | 53,270 | 53,174 | 46,649 | 44,773 | 44,314 | 44,646 | 43,485 | 43,432 | 43,161 | 44,002 | 49,608 | 52,383 |
| 32,528 | 34,697 | 37,578 | 39,676 | 38,907 | 39,921 | 39,189 | 37,750 | 37,777 | 37,842 | 39,304 | 40,608 | 43,174 | 46,078 | 47,480 |
| 15,212 | 15,366 | 16,433 | 16,722 | 16,975 | 16,678 | 16,298 | 16,639 | 16,679 | 16,638 | 17,991 | 19,470 | 20,358 | 21,925 | 22,726 |

**TABLE F-3** Agricultural Research Funding in the Public and Private Sectors, 1970–1998

| FY | Public R&D Funding, thousands of nominal dollars | Private R&D Funding, thousands of nominal dollars |
| --- | --- | --- |
| 1970 | 536,619 | 464,300 |
| 1971 | 581,860 | 487,100 |
| 1972 | 627,100 | 507,400 |
| 1973 | 670,748 | 576,100 |
| 1974 | 729,227 | 669,290 |
| 1975 | 823,521 | 708,540 |
| 1976 | 898,363 | 817,780 |
| 1977 | 1,031,712 | 953,950 |
| 1978 | 1,157,070 | 1,079,109 |
| 1979 | 1,247,217 | 1,204,080 |
| 1980 | 1,367,212 | 1,453,024 |
| 1981 | 1,528,582 | 1,468,190 |
| 1982 | 1,641,571 | 1,651,512 |
| 1983 | 1,703,575 | 1,794,203 |
| 1984 | 1,768,951 | 2,045,965 |
| 1985 | 1,927,991 | 2,167,211 |
| 1986 | 2,014,782 | 2,320,865 |
| 1987 | 2,160,548 | 2,278,197 |
| 1988 | 2,301,184 | 2,571,360 |
| 1989 | 2,445,936 | 2,745,153 |
| 1990 | 2,598,294 | 2,971,347 |
| 1991 | 2,780,467 | 3,172,941 |
| 1992 | 2,913,161 | 3,207,266 |
| 1993 | 2,970,910 | 3,463,213 |
| 1994 | 3,111,548 | 3,556,593 |
| 1995 | 3,168,752 | 3,884,896 |
| 1996 | 3,148,023 | 3,960,789 |
| 1997 | 3,235,060 | 4,381,220 |
| 1998 | 3,403,899 | 4,559,514 |

Note: For information drawn from *Inventory of Agricultural Research* publications, data adjustments were necessary to produce consistent series for period. Data were not available from *Inventory* for 1971. Thus, the figure for 1971 is the average of 1970 and 1972. With respect to more recent years, CRIS stopped publishing *Inventory* in 1997 and moved to a Web-based reporting system. The new data system differs slightly from system used in *Inventory*. Data for 1998 do not include Forest Service's research budget. Constant dollar estimates are based on deflators for research expenditures (see Table F-11).

Source: Public numbers based on USDA Current Research Information System (CRIS), *Inventory of Agricultural Research*, various years; http://www.ers.usda.gov/data/agresearchfunding/; private numbers based on Klotz et al. (1995).

| Public R&D Funding, thousands of nominal dollars | Private R&D Funding, thousands of nominal dollars | Total Agricultural R&D, thousands of 2000 dollars |
| --- | --- | --- |
| 2,450,316 | 2,120,091 | 4,570,408 |
| 2,552,016 | 2,136,404 | 4,688,419 |
| 2,668,510 | 2,159,149 | 4,827,659 |
| 2,704,627 | 2,322,984 | 5,027,611 |
| 2,710,881 | 2,488,067 | 5,198,948 |
| 2,820,277 | 2,426,507 | 5,246,784 |
| 2,916,762 | 2,655,130 | 5,571,892 |
| 3,174,498 | 2,935,231 | 6,109,729 |
| 3,315,386 | 3,092,003 | 6,407,389 |
| 3,282,151 | 3,168,632 | 6,450,783 |
| 3,216,969 | 3,418,880 | 6,635,850 |
| 3,273,194 | 3,143,876 | 6,417,071 |
| 3,302,960 | 3,322,961 | 6,625,921 |
| 3,307,912 | 3,483,890 | 6,791,802 |
| 3,228,012 | 3,733,513 | 6,961,525 |
| 3,341,406 | 3,755,998 | 7,097,404 |
| 3,357,970 | 3,868,108 | 7,226,077 |
| 3,462,416 | 3,650,957 | 7,113,373 |
| 3,491,933 | 3,901,912 | 7,393,845 |
| 3,504,206 | 3,932,884 | 7,437,090 |
| 3,539,910 | 4,048,157 | 7,588,068 |
| 3,682,737 | 4,202,570 | 7,885,308 |
| 3,763,773 | 4,143,754 | 7,907,528 |
| 3,751,149 | 4,372,743 | 8,123,892 |
| 3,789,948 | 4,332,025 | 8,121,974 |
| 3,750,002 | 4,597,510 | 8,347,512 |
| 3,626,754 | 4,563,121 | 8,189,876 |
| 3,606,533 | 4,884,303 | 8,490,836 |
| 3,648,338 | 4,886,939 | 8,535,276 |

**TABLE F-4** Amount and Distribution of Major Sources of Revenues of US State Agricultural Experiment Stations, 1980–2000

| Sources | Revenue, millions of current dollars | | |
|---|---|---|---|
| | 1980 | 1990 | 2000 |
| Regular federal appropriations | 136.9 | 223.6 | 292.6 |
|   Hatch, regional research, and other nongrant funds | 127.2 | 163.9 | 200.9 |
|   CSRS/CSREES special grants | 9.6 | 39.7 | 47.0 |
|   Competitive grants, including NRI | — | 20.0 | 44.7 |
| Other federal government research funds | 91.8 | 193.3 | 360.4 |
|   Contracts, grants, and cooperative agreements with USDA agencies | 24.4 | 49.5 | 75.0 |
|   Contracts, grants, and cooperative agreements with non-USDA federal agencies | 67.4 | 143.9 | 285.4 |
| State government appropriations | 446.9 | 877.9 | 1,117.8 |
| Industry, commodity groups, foundations[b] | 74.0 | 210.0 | 340.9 |
| Other funds (product sales) | 55.2 | 91.6 | 118.0 |
| Total | 804.8 | 1,596.5 | 2,229.7 |

[a]Obtained by deflating data in first three columns by using Huffman and Evenson (1993, pp. 95–97 and updated to 2000) agricultural-research price index with 2000 = 1.00.
[b]Amount received from industry and "other nonfederal sources," excluding state appropriations and product sales or self-generated.

Source: USDA 1982, 1991, 2001 CRIS data.

**TABLE F-5** Sources of Revenue for REE Intramural Research Expenditures, 1980–2000

| Agency | Revenue, millions of current dollars | | |
|---|---|---|---|
| | 1980 | 1990 | 2000 |
| Agricultural Research Service | 360.3 | 580.1 | 794.9 |
| Regular federal appropriations | 360.3 | 570.9 | 775.7 |
| Other funds | 0 | 9.2 | 19.2 |
| Economic Research Service | 42.6 | 51.3 | 72.5[a] |
| Regular federal appropriations | 42.4 | 51.3 | 71.6 |
| Other funds | 0.2 | 0 | 0.9 |
| Total ARS and ERS | 402.9 | 631.4 | 867.4 |
| Regular federal appropriations | 402.7 | 622.2 | — |
| Other funds | 0.2 | 9.2 | — |

[a]Obtained by deflating data in first three columns by using Huffman and Evenson (1993, pp. 95–97 and updated to 2000) agricultural-research price index with 2000 = 1.00.
[b]Data for 1999. ERS did not report any data for CRIS for 2000.

| Revenue, millions of constant 2000 dollars[a] | | | Distribution, % | | |
|---|---|---|---|---|---|
| 1980 | 1990 | 2000 | 1980 | 1990 | 2000 |
| 322.1 | 305.0 | 292.6 | 17.0 | 14.0 | 13.1 |
| 298.8 | 223.6 | 200.9 | [15.8] | [10.3] | [9.0] |
| 22.6 | 54.2 | 47.0 | [1.2] | [2.5] | [2.1] |
| — | 27.3 | 44.7 | — | [1.2] | [2.0] |
| 216.0 | 263.7 | 360.4 | 11.4 | 12.1 | 16.2 |
| 57.4 | 67.5 | 75.0 | [3.0] | [3.1] | [3.4] |
| 158.6 | 196.3 | 285.4 | [8.4] | [9.0] | [12.8] |
| 1,051.5 | 1,197.7 | 1,117.8 | 55.5 | 55.0 | 50.1 |
| 174.1 | 286.5 | 340.9 | 9.2 | 13.2 | 15.3 |
| 129.8 | 125.0 | 118.0 | 6.9 | 5.7 | 5.3 |
| 1,893.6 | 2,178.0 | 2,229.7 | 100.0 | 100.0 | 100.0 |

| Revenue, millions of constant 2000 dollars[a] | | |
|---|---|---|
| 1980 | 1990 | 2000 |
| 847.8 | 790.3 | 794.9 |
| 847.8 | 777.8 | 775.7 |
| 0 | 12.5 | 19.2 |
| 100.2 | 69.9 | 74.8[b] |
| 99.8 | 69.9 | 73.9 |
| 0.4 | 0 | 0.9 |
| 948.0 | 860.2 | 869.7 |
| 947.6 | 847.7 | — |
| 0.4 | 12.5 | — |

**TABLE F-6** REE Agency Funding Allocation by Goal, FY 2000

| Goal | Funding, thousands of dollars (%) | | | | |
| --- | --- | --- | --- | --- | --- |
| | ARS | CSREES | ERS | NASS | Total, REE |
| Goal 1: To achieve agricultural production that is highly competitive in the global economy | 121,327 (15.30) | 291,154 (25.74) | 20,550 (31.44) | 63,935 (64.36) | 496,966 (23.79) |
| Goal 2: To provide a safe and secure food and fiber system | 326,716 (41.20) | 228,945 (20.24) | 3,744 (5.73) | 3,950 (3.98) | 563,355 (26.97) |
| Goal 3: To achieve a healthier, more well-nourished population | 72,731 (9.17) | 238,175 (21.06) | 16,144 (24.70) | 0 (0) | 327,050 (15.66) |
| Goal 4: To achieve greater harmony between agriculture and the environment | 125,648 (15.85) | 203,112 (17.96) | 12,092 (18.50) | 4,831 (4.86) | 345,683 (16.55) |
| Goal 5: To enhance economic opportunities and the quality of life among families and communities | 146,550 (18.48) | 169,810 (15.01) | 12,833 (19.63) | 26,616 (26.79) | 355,809 (17.03) |
| Total | 792,972 | 1,131,196 | 65,363 | 99,332 | 2,088,863 |

Source: Agency FY 2001 performance plans.

**TABLE F-7** National Summary USDA, State Agricultural Experiment Stations, and Other Institutions, FY 2000

| Research Problem Area | Allocation of REE Agricultural Research Funds[a] | | Allocation of Public Agricultural Research Funds[b] | |
|---|---|---|---|---|
| | Thousands of Dollars | % | Thousands of Dollars | % |
| Administration | 670 | 0.06 | 1,031 | 0.04 |
| Natural Resources and Environment | 173,802 | 15.24 | 507,072 | 17.49 |
| Plants and Their Systems | 420,591 | 36.87 | 976,032 | 33.68 |
| Animals and Their Systems | 201,307 | 17.65 | 675,705 | 23.32 |
| Engineering and Support Systems | 26,594 | 2.33 | 66,016 | 2.28 |
| Food and Nonfood Products: Development, Processing, Quality, and Delivery | 96,360 | 8.45 | 183,868 | 6.34 |
| Economic Markets and Policy | 30,717 | 2.69 | 132,892 | 4.59 |
| Human Nutrition, Food Safety, and Human Health and Well-Being | 174,077 | 15.26 | 278,796 | 9.62 |
| Family and Community Systems | 6,987 | 0.61 | 40,423 | 1.39 |
| Research Support, Administration, and Communication | 9,524 | 0.83 | 36,221 | 1.25 |
| Total | 1,140,628 | 100.00 | 2,898,056 | 100.00 |

[a]Includes regular appropriations used for inhouse research by USDA research agencies (note that ERS and FS did not report to FY 2000 CRIS) and expenditures of formula and grant funding administered by CSREES and distributed to state agricultural experiment stations and other cooperating institutions; programs included are Hatch, McIntire-Stennis, Evans-Allen, Animal Health, Special Grants, Competitive Grants, Small Business Innovation Research Grants, and other specific grant programs.

[b]Total public funds represent the sum of USDA-appropriated funding, CSREES-administered funding, other funding from USDA, other federal funding, and state appropriations. Not included are expenditures of funds by the state agricultural experiment stations and other cooperating institutions received from sources outside federal government, including sale of products (self-generated), industry grants, and miscellaneous nonfederal sources.

**TABLE F-8a** Research, Education, and Economics by Function, Agency, and Type of Award, FY 1985–2001 Actual and FY 2002 Estimate

Budget Authority in millions of nominal dollars

|      | Research |     |      | CSREES  |             |                |
|------|----------|-----|------|---------|-------------|----------------|
| Year | ARS      | ERS | NASS | Formula | Competitive | Administrative |
| 1985 | 494      | 47  | 8    | 199     | 54          | 45             |
| 1986 | 473      | 44  | 8    | 189     | 49          | 40             |
| 1987 | 521      | 45  | 3    | 189     | 47          | 136            |
| 1988 | 544      | 48  | 4    | 202     | 45          | 105            |
| 1989 | 569      | 50  | 3    | 203     | 40          | 100            |
| 1990 | 593      | 51  | 3    | 203     | 43          | 136            |
| 1991 | 631      | 54  | 3    | 212     | 73          | 160            |
| 1992 | 671      | 59  | 4    | 220     | 98          | 181            |
| 1993 | 672      | 59  | 4    | 220     | 98          | 157            |
| 1994 | 706      | 55  | 4    | 226     | 103         | 158            |
| 1995 | 710      | 54  | 4    | 226     | 101         | 156            |
| 1996 | 701      | 53  | 4    | 222     | 94          | 152            |
| 1997 | 711      | 54  | 3    | 222     | 112         | 155            |
| 1998 | 746      | 72  | 3    | 222     | 97          | 99             |
| 1999 | 795      | 63  | 4    | 237     | 119         | 111            |
| 2000 | 831      | 64  | 4    | 238     | 234         | 114            |
| 2001 | 924      | 68  | 4    | 240     | 252         | 145            |
| 2002 | 1,033    | 70  | 4    | 242     | 143         | 167            |

| Extension and Education | | | | Statistics | Facilities | |
|---|---|---|---|---|---|---|
| | CSREES | | | | | |
| NAL | Formula | Competitive | Administrative | NASS | ARS | Total |
| 12 | 259 | 5 | 87 | 50 | 22 | 1,281 |
| 11 | 247 | 3 | 86 | 48 | 6 | 1,214 |
| 11 | 253 | 3 | 91 | 54 | 1 | 1,354 |
| 12 | 260 | 3 | 93 | 58 | 15 | 1,389 |
| 14 | 261 | 3 | 97 | 61 | 17 | 1,418 |
| 15 | 264 | 4 | 106 | 64 | 13 | 1,494 |
| 17 | 275 | 5 | 127 | 73 | 41 | 1,671 |
| 18 | 287 | 5 | 135 | 79 | 66 | 1,823 |
| 18 | 287 | 5 | 142 | 77 | 35 | 1,774 |
| 18 | 298 | 6 | 141 | 78 | 33 | 1,827 |
| 18 | 298 | 9 | 142 | 77 | 44 | 1,838 |
| 19 | 294 | 9 | 140 | 77 | 30 | 1,796 |
| 19 | 294 | 26 | 141 | 97 | 69 | 1,903 |
| 19 | 294 | 8 | 139 | 115 | 81 | 1,894 |
| 20 | 302 | 8 | 146 | 100 | 56 | 1,962 |
| 20 | 303 | 73 | 128 | 96 | 53 | 2,157 |
| 20 | 304 | 67 | 142 | 97 | 74 | 2,338 |
| 22 | 307 | 28 | 146 | 110 | 192 | 2,464 |

**TABLE F-8b** Research, Education, and Economics by Function, Agency, and Type of Award, FY 1985–2001 Actual and FY 2002 Estimate

Budget Authority, millions of constant 2000 dollars

|      | Research |     |      | CSREES  |             |                |
|------|----------|-----|------|---------|-------------|----------------|
| Year | ARS      | ERS | NASS | Formula | Competitive | Administrative |
| 1985 | 856 | 81 | 14 | 345 | 94  | 78  |
| 1986 | 805 | 73 | 13 | 315 | 82  | 67  |
| 1987 | 835 | 72 | 5  | 303 | 75  | 218 |
| 1988 | 825 | 73 | 6  | 307 | 68  | 159 |
| 1989 | 815 | 72 | 4  | 291 | 57  | 143 |
| 1990 | 808 | 69 | 4  | 277 | 59  | 185 |
| 1991 | 836 | 72 | 4  | 281 | 97  | 212 |
| 1992 | 867 | 76 | 5  | 284 | 127 | 234 |
| 1993 | 848 | 74 | 5  | 278 | 124 | 198 |
| 1994 | 860 | 67 | 5  | 275 | 125 | 192 |
| 1995 | 840 | 64 | 5  | 267 | 120 | 185 |
| 1996 | 808 | 61 | 5  | 256 | 108 | 175 |
| 1997 | 793 | 60 | 3  | 247 | 125 | 173 |
| 1998 | 800 | 77 | 3  | 238 | 104 | 106 |
| 1999 | 820 | 65 | 4  | 245 | 123 | 115 |
| 2000 | 831 | 64 | 4  | 238 | 234 | 114 |
| 2001 | 880 | 65 | 4  | 229 | 240 | 138 |
| 2002 | 958 | 65 | 4  | 224 | 133 | 155 |

Note: Constant dollar estimates based on deflators for research expenditures (see Table F-11).

| | Extension and Education | | | Statistics | Facilities | |
|---|---|---|---|---|---|---|
| | CSREES | | | | | |
| NAL | Formula | Competitive | Administrative | NASS | ARS | Total |
| 21 | 449 | 9 | 151 | 87 | 38 | 2,223 |
| 18 | 412 | 5 | 143 | 80 | 10 | 2,023 |
| 18 | 405 | 5 | 146 | 87 | 2 | 2,171 |
| 18 | 395 | 5 | 141 | 88 | 23 | 2,108 |
| 20 | 374 | 4 | 139 | 87 | 24 | 2,030 |
| 20 | 360 | 5 | 144 | 87 | 18 | 2,036 |
| 23 | 364 | 7 | 168 | 97 | 54 | 2,215 |
| 23 | 371 | 6 | 174 | 102 | 85 | 2,354 |
| 23 | 362 | 6 | 179 | 97 | 44 | 2,238 |
| 22 | 363 | 7 | 172 | 95 | 40 | 2,223 |
| 21 | 353 | 11 | 168 | 91 | 52 | 2,177 |
| 22 | 339 | 10 | 161 | 89 | 35 | 2,069 |
| 21 | 328 | 29 | 157 | 108 | 77 | 2,121 |
| 20 | 315 | 9 | 149 | 123 | 87 | 2,031 |
| 21 | 312 | 8 | 151 | 103 | 58 | 2,025 |
| 20 | 303 | 73 | 128 | 96 | 53 | 2,158 |
| 19 | 290 | 64 | 135 | 92 | 70 | 2,226 |
| 20 | 285 | 26 | 135 | 102 | 178 | 2,285 |

**TABLE F-9** ARS Funding of Cooperative Activities, FY 1998–2001

Funding, thousands of nominal dollars

| Cooperative activity | 2001 | 2000 | 1999 | 1998 |
|---|---|---|---|---|
| Cooperative agreements | 91,561 | 76,015 | 67,945 | 51,907 |
| Research support agreements | 20,217 | 23,347 | 21,510 | 23,570 |
| Research contracts | 30 | 36 | 3,559 | 19,461 |
| Grants | 25,214 | 28,942 | 15,912 | 13,103 |
| Total ARS obligations for extramural agreements[a] | 137,022 | 128,340 | 108,926 | 108,041 |
| Total ARS budget authorization[b] | 1,019,000 | 903,000 | 817,000 | 845,000 |
|  | (13%) | (14%) | (13%) | (13%) |
| Extramural pass-through funding[c] | 106,180 | 87,458 | 84,164 | 79,161 |

[a] Does not include extramural pass-through funding, shown in last row.
[b] With percentage of ARS budget.
[c] Pass-through agreements initiated from funding appropriated to ARS.

Source: ARS, 2002.

**TABLE F-10a** Congressional Earmarks for ARS Research and CSREES Research, Education, and Extension, Nominal Dollars

Budget Authority, millions of nominal dollars

| Year | ARS | CSREES[a] | | Total |
| --- | --- | --- | --- | --- |
| | | Research | Extension | |
| 1993 | 0 | 82 | 5 | 87 |
| 1994 | 8 | 81 | 6 | 94 |
| 1995 | 9 | 72 | 7 | 89 |
| 1996 | 6 | 71 | 7 | 84 |
| 1997 | 5 | 70 | 7 | 83 |
| 1998 | 15 | 76 | 6 | 97 |
| 1999 | 20 | 87 | 7 | 114 |
| 2000 | 22 | 86 | 7 | 116 |
| 2001 | 34 | 115 | 13 | 163 |
| 2002 | 55 | 132 | 12 | 199 |

Note: Budget authority means authority provided by law to incur financial obligations that will result in outlays. Sources of budget authority included in this analysis include annual appropriations, supplemental appropriations, trust funds, and mandatory spending under substantive law.

ARS estimates include $17.5 million made available under the Agricultural Risk Protection Act in 2001 and $113 million under Emergency Supplemental to Respond to Terrorism in 2002, consisting of $40 million for research and $73 million for facilities. CSREES estimates for 2002 do not include $20.6 million appropriated in Emergency Supplemental to Office of the Secretary and transferred to CSREES.

[a] Estimates for CSREES earmarks are based on total appropriations for special research grants (including improved pest control) and federal administration, excluding amounts provided for CSREES administrative costs.

**TABLE F-10b** Congressional Earmarks for ARS Research and CSREES Research, Education, and Extension, Constant 2000 Dollars

Budget Authority, millions of constant 2000 dollars

|      |     | CSREES[a] | | |
| --- | --- | --- | --- | --- |
| Year | ARS | Research | Extension | Total |
| 1993 | 0   | 104 | 8  | 112 |
| 1994 | 10  | 99  | 7  | 116 |
| 1995 | 11  | 85  | 8  | 104 |
| 1996 | 7   | 82  | 8  | 97  |
| 1997 | 6   | 78  | 8  | 92  |
| 1998 | 16  | 82  | 6  | 104 |
| 1999 | 21  | 90  | 7  | 118 |
| 2000 | 22  | 86  | 7  | 116 |
| 2001 | 32  | 110 | 12 | 154 |
| 2002 | 51  | 122 | 11 | 184 |

Note: Constant dollar estimates based on deflators for research expenditures (see Table F-11).
Budget authority means authority provided by law to incur financial obligations that will result in outlays. Sources of budget authority included in this analysis include annual appropriations, supplemental appropriations, trust funds, and mandatory spending under substantive law.
ARS estimates include $17.5 million made available under the Agricultural Risk Protection Act in 2001 and $113 million under Emergency Supplemental to Respond to Terrorism in 2002, consisting of $40 million for research and $73 million for facilities. CSREES estimates for 2002 do not include $20.6 million appropriated in Emergency Supplemental to Office of the Secretary and transferred to CSREES.

[a] Estimates for CSREES earmarks are based on total appropriations for special research grants (including improved pest control) and federal administration, excluding amounts provided for CSREES administrative costs.

**TABLE F-11** Price Index for Research, 2000 Constant Dollar R&D Deflators

| Year | Deflator | Year | Deflator | Year | Deflator |
|---|---|---|---|---|---|
| 1970 | 0.219 | 1981 | 0.467 | 1992 | 0.774 |
| 1971 | 0.228 | 1982 | 0.497 | 1993 | 0.792 |
| 1972 | 0.235 | 1983 | 0.515 | 1994 | 0.821 |
| 1973 | 0.248 | 1984 | 0.548 | 1995 | 0.845 |
| 1974 | 0.269 | 1985 | 0.577 | 1996 | 0.868 |
| 1975 | 0.292 | 1986 | 0.600 | 1997 | 0.897 |
| 1976 | 0.308 | 1987 | 0.624 | 1998 | 0.933 |
| 1977 | 0.325 | 1988 | 0.659 | 1999 | 0.969 |
| 1978 | 0.349 | 1989 | 0.698 | 2000 | 1 |
| 1979 | 0.38 | 1990 | 0.734 | 2001 | 1.05 |
| 1980 | 0.425 | 1991 | 0.755 | 2002[a] | 1.078 |
|  |  |  |  | 2003[a] | 1.122 |

Notes: Research price index reflects cost of doing research in public universities over time, holding composition of inputs constant; 70% of weight is for faculty compensation (salary and benefits), and 30% for nonfaculty inputs. Faculty salary component is weighted average of salaries of faculty at assistant, associate, and full professor ranks. The cost of nonfaculty inputs is approximated by implicit price deflator for state and local government expenditures (of the Bureau of Economic Analysis).

[a]The value for this year is an estimate.

Source: Huffman (2002).

# APPENDIX G

# REE Dissemination And Outreach Efforts

The US Department of Agriculture (USDA) Research, Education, and Economics (REE) mission area uses multiple outlets for disseminating news, general information, services, and products, such as Web sites, databases, newsletters, and reports. An integral part of this diffusion has been the mission area's attempts to reach out to its clients and partners. Cooperative Extension—the primary vehicle for technology transfer to users—is a critical element of the dissemination process. Each REE agency offers electronic databases and publications for access by the general public, and all USDA mission areas, including REE, are listed on the USDA "Services and Programs" Web page with links to each agency (USDA, 2002k).

## AGRICULTURAL RESEARCH SERVICE

Among the publications of the Agricultural Research Service (ARS), its monthly magazine, *Agricultural Research*, is available for viewing on the ARS Web site and in paper form. *Agricultural Research* details USDA's scientific research and other newsworthy scientific and agricultural information. It is available on ARS's "News and Information" Web page at no charge, but a fee is required for subscription to the paper form. Through the Web site, visitors have full-text access to *Agricultural Research* dating back to May 1996 and index-only access as far back as September 1978 (USDA, 2002a).

In addition to *Agricultural Research*, the "News and Information" page provides a compilation of updated and archived agricultural news. The committee notes that some ARS research is not always communicated to the public via official agency press releases (see Box G-1). The "News and Information" page also

> **BOX G-1**
> **Is ARS Highlighting Its Most Important Research?**
> **A Missed Opportunity**
>
> On May 15, 2002, a press release titled "Spray Weeds with Vinegar?" appeared on the ARS Web site. The press release highlights research by ARS scientists offering the first scientific evidence that vinegar may be a potent weedkiller that is inexpensive and environmentally safe.
>
> The same day, the *Wall Street Journal* (Chase, 2002) reported on another ARS research study "of healthy women over 50 [that] found that moderate alcohol consumption—one or two drinks a day—can improve their response to insulin and reduce their blood levels of triglycerides, blood fats that boost the risk of developing diabetes. Type 2 diabetes, linked to obesity and sedentary living, is now soaring at epidemic levels, affecting 16 million Americans." Similarly, *USA Today* (Manning, 2002) reported that "one or two drinks of alcohol a day improve insulin sensitivity in older women, who are at increased risk of diabetes after menopause. In a study of 63 postmenopausal, nondiabetic women, researchers at the Human Nutrition Research Center, part of the US Department of Agriculture in Beltsville, MD, found that those who had one or two drinks before bed—they were given grain alcohol mixed with orange juice—each night for eight weeks had better insulin sensitivity and lower levels of triglycerides, a type of fat, than nondrinkers. Earlier studies have found that moderate drinking reduces heart disease risk, researchers say. The new data show a similar effect on diabetes risk."
>
> The study was not reported on the ARS news and information page.

is linked to *Healthy Animals* (an online ARS newsletter addressing affairs relevant to animal health research), the ARS *Quarterly Report* of various research projects, *Food and Nutrition Research Briefs*, and the *Methyl Bromide Alternatives Newsletter*. Other ARS publications can be accessed from the site, some free and some requiring a fee. The "News and Information" page also links users to "Sci4Kids," a Web site for youth that details the type of work done at ARS and includes information and resources for teachers. "Sci4Kids" is available in Spanish. Other ARS locations also have Web sites providing information useful for teachers and students (see Box G-2). From the ARS "Offices and Programs" page, a link is provided to its "Diversity Outreach" site, which details ARS programs and outreach efforts concerning equal opportunity and civil rights (USDA, 2002a). Although other ARS locations and programs can be located through its Web site, not all of them can be found or accessed easily.

> **BOX G-2**
> **The Carl Hayden Bee Research Center**
>
> The Carl Hayden Bee Research Center (CHBRC) is an ARS facility near the campus of the University of Arizona in Tucson, designed to advance crop pollination and to increase the productivity of honeybees. The center's Web site (*http://gears.tucson.ars.ag.gov/center/index.html*) includes a page titled "For the Classroom," which is dedicated to assisting students and teachers in conducting honeybee research and in planning curricula based on investigations of bees. "For the Classroom" provides a number of links to related topics based on popular articles and stories useful for younger audiences. Visitors can view highly detailed images of honeybees and consult a forum of questions and answers discussed by various experts. A set of online activities is available to help explain to students the importance of mathematical modeling as related to the study of honeybees. Links also are provided to other relevant and useful Web sites.

In 1962, the National Agricultural Library (NAL) was designated a national institution, although that was not mandated by law until 1990 (US Congress, 1990). Its origins can be traced to 1862, when it was established as the departmental library for USDA. Since then, NAL has fallen under the auspices of ARS, and it has been charged with serving as both a national and a departmental library. Designed to provide agricultural information to the general public, researchers, academicians, and decision-makers, it is among the world's largest agricultural libraries. NAL also is responsible for coordinating the libraries at USDA field locations and the state land-grant libraries. Its mission includes providing a center for agricultural data at the international level (USDA, 2002h).

A number of electronic resources and information sources are provided through the NAL Web site. Four newsletters are sponsored by the library: *Agricultural Libraries Information Notes* (*ALIN*), *Animal Welfare Information Center* (*AWIC*), *Probe* (for USDA's Plant Genome Program), and *Vignettes* (published by the NAL Agricultural Trade and Marketing Information Center). The NAL Web site is linked to other USDA Web pages designed for students, and links are provided to a number of previous annual reports. The NAL Web site contains links to agricultural information resources and to resources not overseen directly by the library. The library also maintains links to its services, such as document delivery and the interlibrary-loan process (USDA, 2002h).

A substantial portion of the NAL mission requires the use of online databases to provide extensive agricultural information. From the NAL Web site, at least 17 databases can be accessed, not all of which are maintained by the library. Two of the databases—Agricultural Online Access (AGRICOLA) and Agricultural Network Information Center (AgNIC)—merit brief descriptions. AGRICOLA is a database of agricultural literature and reports in bibliographic form. A broad array of agricultural records addressing such topics as the plant, animal, and soil sciences is available. Books and journal articles can be searched online through this database, but full-text references are not available directly; several of the database entries contain Web links to their full-text versions. The entries in AGRICOLA are not necessarily limited to a particular year, and NAL notes that it includes records of materials dating back several centuries. In 1998, AGRICOLA was made available online to the general public at no charge (USDA, 2002h). AgNIC is an online resource that provides access to agricultural information on a number of subjects, including animal and veterinary sciences, economics, environmental sciences, forestry, and government regulations. Users can pick from general categories, which are linked to related categories or subcategories; or categories can be searched by using keywords, and an agricultural thesaurus is available for searching (USDA, 2002h). AgNIC represents a volunteer-based partnership between NAL, a number of land-grant universities, and several institutions that are agriculture-related. Other government units and citizen groups also participate, and the total number of partners is about 40. Those participating agree to provide focused segments of agricultural information for the database. Moreover, various AgNIC projects are available for viewing at its Web site. AgNIC's institutional structure includes an executive board with a secretariat and a framework of rules and procedures (USDA, 2002h).

In 2001, a panel of experts appointed by USDA reviewed the quality and effectiveness of NAL in relation to its stated mission. The central finding of the review was that NAL's present degree of support renders it unable to maintain its responsibility as a national library effectively while serving as USDA's departmental library. Although surveys demonstrated a general sense of approval of the NAL on the part of USDA staff, the panel found deficiencies in light of site examinations, progress reports on NAL, and surveys of other users. Ultimately, the panel determined that NAL has not yet succeeded in fulfilling its dual role (Vanderhoef et al., 2001). The panel noted that, in light of NAL's current shortcomings, it will be important to further its development as an institution. Much of the panel's review focused on the need to expand and enhance the library's electronic databases and resources, including AGRICOLA and AgNIC. For example, the review suggested that AGRICOLA would operate better if it were given the functionality and breadth of databases of the National Library of Medicine. In addition to making budgetary recommendations, the panel called for an increase in NAL staff and a realignment in which the library would be placed directly under the auspices of the secretary or deputy secretary of agriculture. In sum, the review panel called for an increase in general support with the objective of promoting the library's dual role and the fulfillment of its responsibility to all users (Vanderhoef et al., 2001).

## ECONOMIC RESEARCH SERVICE

The Economic Research Service (ERS) relies on electronic media and print publications to disseminate its work and information. One of its chief publications, *Agricultural Outlook*, is published 10 times per year and is available online for viewing at no charge, and on paper with a subscription fee. *Agricultural Outlook* is USDA's primary resource for agricultural and food-price forecasts. It typically includes data addressing agricultural commodities, general information on the US economy, and other economic indicators. The central focus of *Agricultural Outlook* is short-term forecasts of the economy and agriculture, but long-term examinations also are provided. Through the ERS Web site, past issues of *Agricultural Outlook* can be accessed back to 1995, and an index of publications over the last 5 years is available (USDA, 2002f).

In addition to *Agricultural Outlook*, ERS offers *Food Consumption, Prices, and Expenditures, 1970–97*, a statistical bulletin providing historical data on patterns in food consumption and spending. Available for viewing online at no charge, this publication also can be purchased in print. *Agricultural Resources and Environmental Indicators* also can be found at the ERS Web site. This report addresses the state of natural resources used in the agricultural economy, and it depicts various trends concerning their use. The first edition appeared in 1994. Online editions are free, but there is a fee for the print form (USDA, 2002f).

Other ERS products include the magazines *Food Review* and *Rural America*. *Food Review*, published three times per year, studies patterns in food assistance, consumption, and safety; *Rural America* appears four times per year and addresses issues related to demographic change and the use of research as applied to rural banking. Outlook reports provide current and prospective information on commodity supply, demand, and price conditions, and annual yearbooks provide historical data series on acreage, yield, supply, domestic use, foreign trade, and price and topical articles pertinent to understanding the US and global markets. Publications in professional journals also are available online (USDA, 2002f). ERS offers a number of data products, all of which can be accessed on-line, including state fact sheets, agricultural baseline projections, and data on farm income, farm financial management, production, supply, and distribution, and farm employment (USDA, 2002e). "Briefing rooms" also are available for in-depth discussion of selected issues and provide a synthesis of ERS research on specific topics, questions and answers, recommended readings, and data products.

## NATIONAL AGRICULTURAL STATISTICS SERVICE

The National Agricultural Statistics Service (NASS)—USDA's principal supplier of agricultural statistics—provides links to its publications, which generally are available through its Web site at no cost or for purchase on paper. Access is provided to reports on commodities, state-level statistical information, and crop weather; relevant graphic information; a calendar of reports; and NASS's monthly newsletter, which contains statistical highlights. Another publication, *Trends in Agriculture*, draws on statistical information to capture the nature of changes and trends in US agriculture. With respect to agricultural graphics, users can access displays of crop and livestock data. From the NASS website, state agricultural statistics services can be located. NASS also offers a number of files depicting historical agricultural data, and special requests for data can be made, subject to a fee. The NASS Web site can be searched by keyword and by criteria, such as year or crop name (USDA, 2002i).

From the NASS Web site, users can access a page detailing the R&D activities of the agency. It includes links to agricultural data and detailed maps and images pertaining to US agriculture. For example, users can enter queries that will generate downloadable maps, which can be useful for projects based on geographic information systems. The NASS Web site also contains pages that provide tables of information on such topics as land use. In addition to its data tables and maps, NASS maintains an on-line database of "published estimates" that can be accessed by the general public. The database spans national, state, and county data on crops and livestock and provides the number of farms by state. Although the database provides a rather extensive body of information, NASS acknowledges that it remains under construction (USDA, 2002i).

Other sources of information from NASS include its "News and Coming Events" Web page and "NASS Kids." "News and Coming Events" provides statistical information relevant to agriculture in the form of brief mass-media statements. "NASS Kids" is an online educational resource for youths, which seeks to inform them about the type of work that NASS does, especially in the context of statistics. It offers learning tools in the form of games, a glossary, and elementary historical information about the agency. "NASS Kids" also provides a page of information useful for teachers and lesson planning. Like ARS's "Sci4Kids," "NASS Kids" is available in Spanish, although the Spanish version does not contain the full extent of information found in the English version (USDA, 2002i).

The 1997 Census of Agriculture was the first to be conducted under the auspices of NASS. From the NASS Web site, the census can be accessed comprehensively. Various rankings, highlights, and profiles are linked from the Web page of the census. The data provided online span the national, state, and county levels, and data on US territories, such as Guam and Puerto Rico, can be accessed.

"Special studies" also are provided, addressing, for example, the 1998 Census of Aquaculture and the 1998 Census of Horticulture (USDA, 2002i).

## COOPERATIVE STATE RESEARCH, EDUCATION, AND EXTENSION SERVICE

Although many of the dissemination and outreach efforts of the Cooperative State Research, Education, and Extension Service (CSREES) are similar to those of the other three REE agencies, this agency is also USDA's primary technology-transfer arm.

### Extension

The Cooperative Extension System—supported by CSREES, state, and local governments—functions primarily to disseminate research results to farmers and other citizens. The Extension network serves clients in 3,150 counties in the United States. According to CSREES's Office of Extramural Programs, real aggregate federal funding for public extension has declined, from $332 million in 1991 to $280 million in 2000.

Cooperative Extension programs include the well-established 4-H and Youth Development Program, a nonformal education program and organization for youth. The 4-H program is maintained under the auspices of CSREES, and its mission focuses on expanding opportunities for and helping to develop the abilities of culturally diverse children and adults through the building of supportive environments (USDA, 2002g). The 4-H Web site allows users to access information concerning 4-H programs and partners, as well as how to join. Among the other features of the 4-H Web site are community-related program information and a history of the 4-H program (USDA, 2002g).

Various reports have analyzed the land-grant universities and the extension system according to their outreach ability, as well as how these institutions have tried to bolster outreach and increase dissemination. A 1996 National Research Council report provides several recommendations for colleges of agriculture within land-grant universities with respect to extension (NRC, 1996), including a need for greater systematization of data on the results of extension programs, expanding linkages to other federal agencies, and strengthening the research underpinnings of extension, including in nonfarm programs.

A 1999 report, *Returning to Our Roots: The Engaged Institution*, calls for a stronger sense of "engagement" between land-grant and state universities and the communities that they serve (Kellogg Commission, 1999). The concept of engagement emphasizes the need to abandon one-way contact between institutions and communities in favor of greater collaboration and interaction. In addition to strengthening technology transfer to users, engaged institutions are expected to expand opportunities for students to contribute to the extension system. The 1999 report provides a set of guidelines by which engagement can be measured, including the ability of institutions to respond to those whom they serve, the level of deference given to those who work with or are served by such institutions, the need for institutions to remain neutral in their treatment of potentially controversial topics, the degree to which institutions are accessible, the extent to which institutions' purposes are integrated with their duties as facilitators of student training, how well the different actors within institutions are coordinated, and the extent to which institutions are connected with partners that provide vital resources for their missions (Kellogg Commission, 1999). The land-grant university system is expected to serve as a mechanism for "engagement" with those assisted by extension.

In response to the Kellogg Commission's report (1999), the Extension Committee on Organization and Policy (ECOP) formulated a vision of engagement that considers the impact of changing demographics, advances in technology, and social changes that confront contemporary America (ECOP, 2002). They noted that for extension to meet the new challenges of engagement, its workforce must be empowered to alter programs and their delivery. Having more than 3,000 facilities nationwide, extension has tremendous connectivity and has built its reputation by responding to the grass-roots needs of communities. Those efforts will be most effective if extension and its university partners forge effective alliances with public and private agencies and organizations that provide health and human services, commercial or civic evaluation, and private-sector vendors of technical information (ECOP, 2002).

The structure, function, and processes of extension have been changing. Extension is increasingly playing a universitywide role outside colleges of agriculture in many universities—an arrangement that has provided access to a broader array of university resources and expertise and has fostered more multidisciplinary research. Extension is increasingly engaging stakeholders and other users and is responding to more broadly defined problems that go beyond its traditional focus on agricultural production (NRC, 2002).

## Other CSREES Dissemination and Outreach Efforts

Like the other REE agencies, CSREES provides many of its resources online through its Web site. From its "News and Information" page, users can access current information about CSREES, its mass-media releases, and other relevant news. CSREES also offers a newsletter that is archived back to 2000. Another online resource, "Partners on the Web." is a video magazine detailing national research, education, and extension programs in the United States. Using video streaming technology, visitors can access three episodes, the most recent of which dates back to Spring 2000. From the "News and Information" page, users also can obtain application information and other details about the CSREES Fellows Program (USDA, 2002d).

The "News and Information" page contains a link to "Community Supported Agriculture" (CSA), which is a program maintained by CSREES and NAL. CSA represents an effort to link relevant users with databases on sustainable agriculture and with communities of users engaged in cooperative economic associations. In addition to CSA, visitors can use the "News and Information" page to browse information about the CSREES Competitive Grants Program, current requests for proposals, and a calendar of previous and upcoming CSREES events (USDA, 2002d).

The CSREES Web page home provides a link to "Agriculture in the Classroom," a program that helps to develop students' understanding of the relationships among agriculture, the economy, and society. Representatives of farming associations, government, agribusiness, and higher education participate at the state level to implement this program. USDA works to coordinate and facilitate "Agriculture in the Classroom" (USDA, 2002b).

The Sustainable Agriculture Network (SAN) is the communication and outreach arm of USDA's Sustainable Agriculture Research and Education (SARE) program, a competitive-grants program that supports regional sustainable agriculture research and education. SAN is a cooperative effort among academe, government, and other organizations interested in the sharing of information relevant to sustainable agriculture. In 1991, SAN launched an e-mail discussion forum intended to provide responses to questions concerning sustainable agriculture (Sustainable Agriculture Network, 2002). Currently, 900 users subscribe to the forum. SAN also provides information in a variety of formats, such as electronic diskettes and printed materials.

The Experiment Station Committee on Organization and Policy of the National Association of State Universities and Land-Grant Colleges published a report, *A Science Roadmap for Agriculture*, in 2001. The report considers the progress that could be made if new enterprises allowed the US agricultural system to capitalize on innovations arising from basic science, to respond to the internationalization of markets, to improve the status of rural and urban communities, and to engage in environmental protection. The report was written around a number of "challenges" by which its findings are conveyed (ESCOP, 2001).

## CSREES Databases

### Current Research Information System

The Current Research Information System (CRIS) is USDA's documentation and reporting system for current agricultural, food and nutrition, and forestry research. It contains over 30,000 descriptions of current, publicly supported research projects of the USDA agencies, the state agricultural experiment stations (SAESs), the state land-grant universities, state schools of forestry, cooperating schools of veterinary medicine, and USDA grant recipients. The CRIS database is overseen by the Information Systems and Technology Management (ISTM) unit of CSREES. The database includes information on the type of activity being performed, the people performing it, the location of the activity, progress made, anticipated impacts, and publications that have resulted from it. The public can access CRIS online free of charge, but other products and services, such as information requests that can be made to agency staff, are not available to all people and institutions (USDA, 2002c).

**Agricultural Databases for Decision Support (ADDS) Program**

An information program that can be reached through the CSREES Web pages is the Agricultural Database for Decision Support (ADDS) Program, in which CSREES is a partner. ADDS, Inc.—a private, nonprofit corporation—is a Web site and Internet support center that develops, promotes, and delivers educational materials, datasets, software, and other decision-support tools to agricultural producers and others. Other partners include land-grant universities and the private sector (USDA, 2002c).

**Food and Agricultural Education Information System**

The Food and Agricultural Education Information System (FAEIS) is an online database of higher-education statistics spanning human sciences, agriculture, and the food sciences. Drawing on national data from multiple government agencies, land-grant universities, professional associations, and other databases, FAEIS includes information on renewable natural resources, forestry, general agriculture, and veterinary medicine. It is operated through Texas A&M University (USDA, 2002c).

**Science and Education Impact Databases**

The Science and Education Impact Databases provide information obtained annually from institutions in the land-grant–USDA partnership on the impacts of research, teaching, and extension programs. The databases can be queried by topic, term, and state, and they can be viewed through topical summaries and fact sheets (USDA, 2002c).

**Research Management Information System**

REE research activities are tracked by the Research Management Information System (RMIS), a computer-based documentation and reporting system for current and recently completed CRIS projects in agriculture, food and nutrition, and forestry research. RMIS is designed to provide access to information about research conducted primarily in the REE agricultural research system. Projects cataloged are conducted or sponsored by USDA research agencies, SAESs, the state land-grant university system, other cooperating state institutions, and participants in USDA's NRI Competitive Grants Program. RMIS also tracks patents and CRADAs.

**Research, Education, and Economics Information System**

The 1996 Federal Agriculture Improvement and Reform Act (US Congress, 1996) permitted the construction of an information system that would track and assess affairs in agricultural research and extension. CSREES was charged with bringing together the other REE agencies in an effort to design and put into practice such a system, called the Research, Education, and Economics Information System (REEIS). The impetus for REEIS came from a deficit in the body of REE electronic information concerning the programs that it conducts with its partners—namely universities and other institutions of higher education. Furthermore, the Government Performance and Results Act has required standards for reporting on the status of USDA projects (USDA, 2002j).

REEIS is expected to provide the public with access to information about research results and new technologies while decreasing redundancies in these efforts. It also is intended to create links between similar programs, to harmonize information about REE programs, to meet standards for fiscal responsibility, and to monitor the progress of technologies used in research, economics, extension or education activities. The broader goal of REEIS is to interconnect several databases used by extension and other REE agencies. The Science and Education Resources Development division of CSREES is charged with oversight of REEIS (USDA, 2002j). Although the public has on-line access to minutes of the REEIS National Steering Committee meetings, the future of the REEIS system is unclear (USDA, 2002j).

## SUMMARY

The Research, Education, and Economics mission area disseminates its information and services through a number of channels used by its four agencies. Closely related to dissemination is the mission area's effort to increase its degree of outreach, which in turn requires greater engagement with communities and other users of agricultural technologies, innovations, and education programs. All four of the REE agencies rely on electronic media to disseminate their research and services, and each uses print materials as well. However, collectively and individually, the agencies tend to stress the utility of electronic media for helping to fulfill their mission statements. This effort includes the further development of agency Web pages and online databases available to the general public. Nevertheless, as various reports and user surveys have indicated, not all of REE's electronic resources are well interfaced, and Web sites related to the mission area's work are not entirely accessible from the agency pages.

## REFERENCES

Chase, M. 2002. Drinking alcohol may reduce diabetes risk in middle-aged. The Wall Street Journal. May 15, p. D6.

ECOP (Extension Committee on Organization and Policy). 2002. The Extension System: A Vision for the 21st Century. February. Washington, DC: National Association of State Universities and Land Grant Colleges.

ESCOP (Experiment Station Committee on Organization and Policy). 2001. A Science Roadmap for Agriculture. Washington, DC: Task Force on Building a Science Roadmap for Agriculture, National Association of State Universities and Land-Grant Colleges. Available online at http://www.nasulgc.org/comm_food.htm.

Kellogg Commission on the Future of State and Land-Grant Universities. 1999. Returning to Our Roots: The Engaged Institution. Washington, DC: National Association of State Universities and Land-Grant Colleges. Available online at http://www.nasulgc.org/publications/Kellogg/engage.pdf.

Manning, A. 2002. Researchers weigh in on soaring diabetes rates. USA Today. May 15.

NRC (National Research Council). 1996. Colleges of Agriculture at the Land Grant Universities: Public Service and Public Policy. Washington, DC: National Academy Press.

NRC (National Research Council). 2002. Publicly Funded Agricultural Research and the Changing Structure of US Agriculture. Washington, DC: National Academy Press.

Sustainable Agriculture Network. 2002. Sustainable Agriculture Network. Available online at http://www.sare.org/htdocs/docs/about.html.

US Congress. 1990. P.L. (Public Law) 101-624. Food, Agriculture, Conservation, and Trade Act of 1990.

US Congress. 1996. P.L. (Public Law) 104-127. Federal Agriculture Improvement and Reform Act (FAIR) of 1996.

USDA (US Department of Agriculture). 2002a. ARS News and Information. Washington, DC: Agricultural Research Service, US Department of Agriculture. Available online at http://www.ars.usda.gov/is/.

USDA (US Department of Agriculture). 2002b. Cooperative State Research, Education, and Extension Service. Washington, DC: Cooperative State Research, Education, and Extension Service, US Department of Agriculture. Available online at http://www.reeusda.gov/.

USDA (US Department of Agriculture). 2002c. Cooperative State Research, Education, and Extension Service: CSREES Databases. Washington, DC: Cooperative State Research, Education, and Extension Service, US Department of Agriculture. Available online at http://www.reeusda.gov/1700/programs/database.htm.

USDA (US Department of Agriculture). 2002d. Cooperative State Research, Education, and Extension Service: News and Information. Washington, DC: Cooperative State Research, Education, and Extension Service, US Department of Agriculture. Available online at http://www.reeusda.gov/1700/whatnew/newsinfo.htm.

USDA (US Department of Agriculture). 2002e. Economic Research Service. Washington, DC: Economic Research Service, US Department of Agriculture. Available online at http://www.ers.usda.gov/.

USDA (US Department of Agriculture). 2002f. Economic Research Service Publications. Washington, DC: Economic Research Service, US Department of Agriculture. Available online at http://www.ers.usda.gov/Publications/.

USDA (US Department of Agriculture). 2002g. National 4-H Headquarters. Available online at *http://www.4h-usa.org/*.
USDA (US Department of Agriculture). 2002h. National Agricultural Library. Washington, DC: National Agricultural Library, US Department of Agriculture. Available online at *http://www.nal.usda.gov*.
USDA (US Department of Agriculture). 2002i. National Agricultural Statistics Service. National Agricultural Statistics Service, US Department of Agriculture. Available online at *http://www.usda.gov/nass/*.
USDA (US Department of Agriculture). 2002j. Research, Education, and Economics Information System: Overview. Washington, DC: US Department of Agriculture. Available online at *http://www.reeusda.gov/ree/reeis/reeover.htm*.
USDA (US Department of Agriculture). 2002k. Research, Education, and Economics Mission Area. Washington, DC: US Department of Agriculture. Available online at *http://www.reeusda.gov/ree/*.
Vanderhoef, L., M.A. Apple, K.J. Coulter, W.B. Delauder, J. Hirschman, A. Hoover, P. Hudson, B. Hutchinson, P. Kaufman, M.B. Krewson, P.S. Reed, W. Tabb, and R. Willard. 2001. Report on the National Agricultural Library–2001. Washington, DC: National Agricultural Library, US Department of Agriculture.

# About the Authors

**Laurian J. Unnevehr, University of Illinois, Urbana-Champaign, IL,** *Chair, March–December 2002*

Dr. Unnevehr is a professor in the Department of Agricultural and Consumer Economics, University of Illinois at Urbana-Champaign, where she has been a member of the faculty since 1985. Her research focuses on the social-welfare implications of food safety and diet–health linkages. She spent 1993–1995 on leave at the US Department of Agriculture Economic Research Service. She received a BA in economics from the University of California, Davis and MA and PhD in food economics from Stanford University. Dr. Unnevehr is a member of the Editorial Board of *Food Policy* and the Editorial Council of the *Review of Agricultural Economics*.

**Franklin M. Loew, Becker College, Worcester, MA,** *Chair, July 2000– March 2002*

Dr. Loew is president of Becker College in Worcester, MA. Before joining Becker College, Dr. Loew was president and chief executive officer of Medical Foods, Inc. (1997–1998), where he served as chairman of the board and chairman of the Scientific Advisory Board. Dr. Loew's previous positions include: professor and dean of veterinary medicine at Cornell University; dean of veterinary medicine and chairman of the Department of Environmental Studies at Tufts University; director of the Division of Comparative Medicine at Johns Hopkins University School of Medicine; and professor of physiology, director of the Research Programs in Toxicology, and director of the Animal Resources Centre at the University of Saskatchewan, Canada, where he was also a Medical Research Council fellow. Currently, Dr. Loew has appointments as a visiting scientist at Massachusetts Institute of Technology and as a senior fellow at Tufts University. In 1977, Dr. Loew received the Queen Elizabeth II Jubilee Medal from the governor general of Canada for his work on animal welfare. In 1992, Loew was elected to the Institute of Medicine. He was a member of the National Research Council Board on Biology and Commission on Life Sciences and chair of the Council of the National Research Council Institute of Laboratory Animal Resources. He served on the National Research Council Committee on Improved Models for Toxicity Testing for Human Health Hazard Assessment and the Panel on Animal Health and Veterinary Medicine. He received a BS and a DVM from Cornell University and a PhD in nutrition from the University of Saskatchewan, Canada.

**Ransom Lee Baldwin, University of California, Davis, CA**

Dr. Baldwin is professor of animal science in the Department of Animal Science at the University of California, Davis. His research focuses on nutritional energetics, physiology of lactation, modeling of ruminant digestion and metabolism, and resource use in livestock production. He served on several National Research Council committees, including the Committee on Land-Grant Colleges of Agriculture and the Committee to Revise the Guide for the Care and Use of Laboratory Animals. Dr. Baldwin chaired the National Research Council Subcommittee on Input-Output Relationships in Animal Production. Currently, he serves on the National Research Council Committee on Cost of and Payment for Animal Research. In 1993, Baldwin was elected to the National Academy of Sciences. He received a BS in animal industries from the University of Connecticut and both an MS in dairy nutrition and a PhD in biochemistry and nutrition from Michigan State University.

**Roger N. Beachy, Donald Danforth Plant Science Center, St. Louis, MO**

Dr. Beachy is president of the Donald Danforth Plant Sciences Center in St. Louis, MO. He has expertise in plant biology and biotechnology and in crop production. His research focuses on virology, gene expression in plants, and the protection of crops from viruses by incorporation of viral genes into the plant genome. Dr. Beachy has headed the Division of Plant Biology at The Scripps Research Institute in La Jolla, California, where he was also codirector of the International Laboratory for Tropical Agricultural Biotechnology. Earlier, he was professor and head of the Center for Plant Science and Biotechnology at Washington University. His work at Washington University, in collaboration with Monsanto Co., led to development of the world's first genetically altered food crop, a variety of tomato that was modified for resistance to viral disease. He was elected to the National Academy of Sciences in 1997, received the Wolf

Prize in Agriculture in 2001, and received the Dennis R. Hoagland Award from the American Society of Plant Biologists in 2000. Dr. Beachy serves on the board of directors for ICRISAT, a member of the Consultative Group on International Agricultural Research, and has been a consultant for several companies. He has served on many National Research Council committees, including the Committee on Biobased Industrial Products: National Research and Commercialization Priorities and the Committee on Biological Pest and Pathogen Control. He received a BA in biology from Goshen College in Goshen, Indiana, and a PhD in plant pathology from Michigan State University.

## Carolyn Branch Brooks, University of Maryland Eastern Shore, Princess Anne, MD

Dr. Brooks is dean of the School of Agricultural and Natural Sciences and research director of Land-Grant Programs at the University of Maryland Eastern Shore (UMES). She has expertise in molecular biology and agricultural microbiology, the system of historically black colleges and universities, and university administration. Dr. Brooks has also served UMES in various other capacities, including executive assistant to the president and chief of staff, interim chair of the Department of Agriculture, program coordinator of the Plant and Soil Science Group, codirector of the Center for Plant and Microbial Biotechnology, and director of the Scientific Enrichment Program for Undergraduate Minority Students. Dr. Brooks is a recipient of many professional awards and recognitions, including the First Annual White House Initiative for Historically Black Colleges and Universities Faculty Award for Excellence in Science and Technology and the Outstanding Educator Award from the Maryland Association of Higher Education. She has been an active participant in many professional activities, including the Minority Education Committee of the American Society of Microbiology's Board of Education and Training and numerous US Department of Agriculture and US Agency for International Development review panels. She is chair of the Association of Research Directors of the 1890 institutions and also represents the 1890 land-grant system on the National Association of State Universities and Land-Grant Colleges' Board of Agriculture Assembly's Policy Board of Directors. In 1993, Dr. Brooks served on the National Research Council Panel for Review of Agricultural Sciences Research Proposals Under the US Agency for International Development's Research Grants Program for historically black colleges and universities; and in 1995, she was appointed site review coordinator of the National Research Council Committee on Undergraduate Science Education. Dr. Brooks received her BS and MS in biology from Tuskegee University and a PhD in microbiology from Ohio State University.

## Elizabeth Chornesky, Carmel, CA

Dr. Chornesky is a freelance consultant and a research associate at the University of California, Santa Cruz. She has broad science and policy experience in biodiversity conservation, natural-resource management, invasive species, marine ecosystems, and pesticide alternatives. Dr. Chornesky was director of conservation research at The Nature Conservancy while working on this report. Previously, she was director of stewardship at The Nature Conservancy, a project director and senior analyst at the US Congress Office of Technology Assessment, a research faculty member at Lehigh University, and a research associate at the Smithsonian Institution. She has served on several national advisory committees and as a consultant to organizations, including the National Research Council, the Environmental Protection Agency, the Union of Concerned Scientists, and the Henry A. Wallace Institute for Alternative Agriculture. Dr. Chornesky holds a BA in biology from Cornell University and a PhD in biology from the University of Texas at Austin.

## Edward A. Hiler, Texas A&M University System, College Station, TX

Dr. Hiler is vice chancellor for agriculture and life sciences of the Texas A&M University System and dean of the College of Agriculture and Life Sciences. He is also director of the Texas Agricultural Experiment Station and of the Texas Cooperative Extension. His expertise includes agricultural engineering, soil and water conservation engineering, bioenergy resources, and administration of education and research programs. Dr. Hiler's previous positions include head of the Department of Agricultural Engineering, deputy chancellor for academic and research programs, and deputy chancellor for academic program planning and research. Dr. Hiler also consults for private firms on designing irrigation systems. He is on the Board of Directors of the Riley Memorial Foundation and the Board of CNH Global. He served as president of the American Society of Agricultural Engineers in 1991–1992. In 1987, he was elected to the National Academy of Engineering. He received a BS, an MS, and a PhD in agricultural engineering from the Ohio State University.

## Wallace Edgar Huffman, Iowa State University, Ames, IA

Dr. Huffman is C.F. Curtiss Distinguished Professor of Agriculture and professor of economics and agricultural economics at Iowa State University, Ames, and a fellow of the American Association of Agricultural Economics. His research focuses on R&D management and policy, human capital for agriculture, agricultural productivity and technical change analysis, adoption of technology, agricultural household models, and migration and immigration. Dr. Huffman has been a member of many US Department of Agriculture and state agricultural experiment station research committees. He testified before US House and Senate committees on issues related to agricultural research, education, technology transfer, and the agricultural labor and commodity market. He serves as a reviewer for many professional journals, including the *American Journal of Agricultural Economics,* the *American Economic Review,* the *Review of Economics and Statistics,* and the *Journal of Agricultural and Resource Economics.* He was a member of the National Research Council Committee on Evaluation of Trends in Competency Needs in Agricultural Research at the Doctoral & Post-Doctoral Personnel Level (1983–1988). Dr. Huffman received a BS in agriculture from Iowa State University, and an MA and PhD in economics from the University of Chicago.

**Lonnie J. King, Michigan State University, East Lansing, MI**

Dr. King is dean of the College of Veterinary Medicine at Michigan State University. He has extensive expertise in veterinary medicine and food safety and administrative experience in government and university systems. Previous positions he has held include administrator of the US Department of Agriculture Animal and Plant Health Inspection Service (APHIS) in Washington, DC, and associate administrator and deputy administrator for veterinary services in APHIS. Before his government career, Dr. King was in private practice. His other experience includes work as a field veterinary medical officer, station epidemiologist, and staff member in emergency programs and animal health information. Dr. King has also directed the American Veterinary Medical Association Office of Governmental Relations and is certified in the American College of Veterinary Preventive Medicine. In 1998, he was a member of the National Research Council Committee on Ensuring Safe Food from Production to Consumption. He serves as president of the American Association of Veterinary Medicine Colleges, cochair of the National Alliance for Food Safety, cochair of the National Commission on Veterinary Economic Issues, and lead dean at Michigan State University for food safety with responsibility for the National Food Safety and Toxicology Center. Dr. King received his BS and DVM from Ohio State University and his MS in epidemiology from the University of Minnesota. He has also attended the Senior Executive Program at Howard University and received a master's degree in public administration from American University.

**Larry Kuzminski, Duxbury, MA**

Dr. Kuzminski is retired from Ocean Spray Cranberries, Inc., where he held the position of vice president of technology, research and development, and technology/operations. He currently works as a consultant in food science and technology, technology and operations strategy, food safety, quality systems, and agricultural research that integrates crop production into product and consumer benefit. Dr. Kuzminski has held various positions with the Kellogg Company, including director of food research with Kellogg US and as senior vice president of science and quality, director of noncereal manufacturing, and senior scientist with Kellogg Canada. He is a past president of the Riley Memorial Foundation and has served on numerous boards and advisory committees, including current service on the Food Advisory Committee to the Food and Drug Administration. Dr. Kuzminski received his BA and MA in food chemistry from the University of Toronto, his PhD in food science from the University of Massachusetts, and his MBA from Western Michigan University.

**William B. Lacy, University of California, Davis, CA**

Dr. Lacy is vice provost of University Outreach and International Programs and professor of sociology in the Department of Human and Community Development at the University of California, Davis. He has conducted extensive research on the sociology of science and agriculture. Dr. Lacy was director of the Cornell Cooperative Extension and associate dean at Cornell University and assistant dean for research and assistant director of the experiment station, Pennsylvania State University. He is a fellow of the American Association for the Advancement of Science and past president of the Rural Sociological Society. He received his BS from Cornell University and his MA and PhD from the University of Michigan.

**Thomas L. Lyon, Cooperative Resources International, Shawano, WI**

Mr. Lyon is chief executive officer at Cooperative Resources International, Shawano, WI. He has broad experience in dairy cattle breeding and production, farm operations, and business management. Before his current position, Mr. Lyon was general manager at 21st Century Genetics, where he also held positions in marketing and public relations. Mr. Lyon has undertaken many professional activities in cooperatives, dairy industry, government, universities, and the local community in many capacities, including president of the Wisconsin Federation of Cooperatives, board chairman of the National Cooperative Business Association, member of the National Rural Development Task Force & Co-op 2000 Committee, president of the National Association of Animal Breeders, member of four Wisconsin gubernatorial commissions, member of the Executive Committee of the National Agricultural, Research, Extension, Education, and Economics Advisory Board, and president of the Board of Regents of the University of Wisconsin System. Mr. Lyon received his BS in dairy science from Iowa State University.

## Kristen W. McNutt, Consumer Choices, Inc., Santa Cruz, CA

Dr. McNutt is president of Consumer Choices, Inc. and editor of *Consumer Magazines Digest*. She provides consulting services to public- and private-sector clients on consumer communication on food-related health topics. She occasionally writes the editorial column "In the Consumer Interest" for *Nutrition Today* and previously wrote "A View from America" for the *British Nutrition Foundation Bulletin*. Dr. McNutt is also associate editor of the *Encyclopedia of Food Science and Technology*. She served on the National Institute of Diabetes and Digestive and Kidney Diseases Advisory Council and was a member of the 1998 National Institutes of Health Office of Alternative Medicine Special Emphasis Panel. She has also served on the Food and Drug Administration Food Advisory Committee. Dr. McNutt has been a member of the National Research Council Committee on Technological Options to Improve Nutritional Attributes of Animal Products, the Cooperative Program for NAS/ASRT (Egypt), and most recently the National Research Council Subcommittee on Drug Use in Food Animals. Dr. McNutt received a BA in chemistry from Duke University, an MS in nutrition from Columbia University College of Physicians & Surgeons, and a PhD in biochemistry from Vanderbilt University. She also received a JD from DePaul College of Law.

## William L. Ogren, Hilton Head Island, SC

Dr. Ogren is retired research leader in the Photosynthesis Research Unit of the Agricultural Research Service (ARS) of the US Department of Agriculture. He has broad experience in research program administration, biochemistry, the physiology and genetics of photosynthesis and photorespiration, crop physiology, crop production, and plant science. He has been a member of the National Academy of Sciences since 1986. Dr. Ogren's professional experience includes plant physiologist at ARS and professor in the Department of Agronomy, University of Illinois. He received many professional honors and awards, including the Crop Science Award from the Crop Science Society of America, the Charles F. Kettering Award for excellence in photosynthesis research from the American Society of Plant Biologists, the US Department of Agriculture Superior Service Award, and the Alexander von Humboldt Medal for contributions to American agriculture; he is a fellow of the American Society of Agronomy and the American Academy of Arts and Sciences. Dr. Ogren received a BS in chemistry from the University of Wisconsin, Madison, and a PhD in biochemistry from Wayne State University.

### David Pimentel, Cornell University, Ithaca, NY

Dr. Pimentel is professor of ecology and agricultural sciences in the Department of Entomology, Cornell University. He has broad expertise in ecology and in ecological and economic aspects of agricultural sciences. His past positions include professor and head of the Department of Entomology and Limnology at Cornell University and chief of the Tropical Research Laboratory, US Public Health Service in Puerto Rico. Dr. Pimentel served on many National Research Council study committees and panels, chaired the Environmental Studies Board and the Board of Science and Technology for International Development, and was a member of the Committee on the Role of Alternative Farming Methods in Modern Production Agriculture. He received a BS from the University of Massachusetts and a PhD from Cornell University and was an Organization for European Economic Cooperation fellow at Oxford University.

### Robert J. Reginato, Chandler, AZ

Dr. Reginato has extensive experience in theoretical and experimental soil science and remote-sensing techniques to assess crop stress. He also has broad experience in research-program administration and outreach activities. He has held positions at the US Department of Agriculture Agricultural Research Service including associate administrator, director of the Pacific West Area, and research leader of the Phoenix, AZ, Soil-Plant-Atmosphere Research Unit. Dr. Reginato served as the interim director of the statewide Sustainable Agriculture Research and Education Program for the University of California. He received a BS in soil management from the University of California at Davis, an MS in agronomy from the University of Illinois, and a PhD in soil science from the University of California, Riverside.

### John W. Suttie, University of Wisconsin, Madison, WI

Dr. Suttie is a retired professor of biochemistry and nutritional sciences at the University of Wisconsin, Madison. He has broad expertise in biochemistry and human nutrition. His research activities are directed toward the metabolism, mechanism of action, and nutritional significance of vitamin K. Dr. Suttie has served as president of the American Society for Nutritional Sciences. He has also served as president of the Federation of American Societies for Experimental Biology. In 1996, Dr. Suttie was elected to the National Academy of Sciences, and he is a past member of the National Research Council's Board on Agriculture and Natural Resources. Dr. Suttie received his BS, MS, and PhD from the University of Wisconsin, Madison.

# About the Subcommittees

Three subcommittees, the Subcommittee on Environmental Quality and Natural Resources, the Subcommittee on Food and Health, and the Subcommittee on Economic and Social Development in a Global Context, generated white papers that provided input into the synthesis committee's final report. Members of the subcommittees also were extremely helpful in providing the synthesis committee with other data and written materials throughout the study.

Although sections of the white papers and other materials authored by subcommittee members were used in the preparation of this report, the report as a whole represents a consensus of the synthesis committee only.

### Subcommittee on Environmental Quality and Natural Resources

**G. Philip Robertson, W.K. Kellogg Biological Station, Michigan State University, Hickory Corner, MI,** *Chair*

Dr. Robertson has been a professor of crop and soil sciences at Michigan State University since 1985 and director of the National Science Foundation (NSF) Long-Term Ecological Research Program in Agricultural Ecology at the W.K. Kellogg Biological Station since 1988. His research interests include nitrogen biogeochemistry and in particular nitrogen conservation in field-crop ecosystems, biogenic sources of atmospheric trace-gas fluxes, and the functional significance of soil microbial diversity. Dr. Robertson has been a postdoctoral fellow at the Royal Swedish Academy of Sciences (1980–1981) and a sabbatical scholar at Cooperative Research Centres in Adelaide (1993–1994) and Brisbane (2001–2002), Australia. His service includes memberships on various grant panels at the US Department of Agriculture (USDA) and NSF and directorship of the USDA Fund for Rural America Environment Program in 1997–1998. Dr. Robertson also served on the National Research Council Committee on an Evaluation of the US Department of Agriculture National Research Initiative Competitive Grants Program (1998–1999). He received a BA from Hampshire College and a PhD in ecology and evolutionary biology from Indiana University.

**Janet C. Broome, Sustainable Agriculture Research and Education Program, University of California, Davis, CA**

Dr. Broome is the associate director of the University of California's statewide Sustainable Agriculture Research and Education Program (UC SAREP) based in Davis. She works with the director to provide leadership and administrative support for the first sustainable-agriculture program at a land-grant university. She leads the agricultural-chemical use-risk reduction program and the Biologically Integrated Farming Systems (BIFS) competitive-grants program, and she is the lead scientist for a $1 million grants program for the development of alternatives to methyl bromide. She conducts research in ecologically based pest management, working in weather-driven disease-forecasting models. In addition, working with the UC integrated pest management project, she developed a plant-disease model database for the PestCast statewide weather-monitoring and disease-forecasting network. Dr. Broome served as an environmental research scientist for the California Environmental Protection Agency Department of Pesticide Regulation in 1994–1997. She has provided expert testimony to the California state legislature on the implementation of the Food Quality Protection Act. Dr. Broome has broad expertise in ecology, epidemiology and control of fungal plant pathogens, sustainable viticulture, integrated farming systems, and ecologic pest management. She also has extensive understanding of pesticide-related regulatory issues. Dr. Broome received a BA in biologic sciences from Swarthmore College and a PhD in plant pathology from the University of California, Davis.

**Elizabeth Chornesky, The Nature Conservancy, Santa Cruz, CA\***

See About the Authors.

**Jane Frankenberger, Department of Agricultural and Biological Engineering, Purdue University, West Lafayette, IN**

Dr. Frankenberger is associate professor in the Department of Agricultural and Biological Engineering at Purdue University. She leads statewide extension

---

\* Synthesis Committee member.

and research programs in soil and water engineering, watershed assessment and management, geographic information systems, agricultural drainage, wellhead protection, and watershed modeling. Dr. Frankenberger has also worked in international development in Senegal (1984–1990) and the Democratic Republic of Congo (1979–1982). Dr. Frankenberger received a BA in physics from St. Olaf College, an MS in agricultural engineering from the University of Minnesota, and a PhD in agricultural and biologic engineering from Cornell University.

**Paul Johnson, Oneota Slopes Farm, Decorah, IA**

Dr. Johnson and his family have owned and operated Oneota Slopes Farm near Decorah, IA, since 1974. Their operation has involved dairy, corn, soybeans, hay, beef cattle, sheep, and Christmas trees. Dr. Johnson served three terms in the Iowa State Legislature, 1984–1990, and was chief of the Soil Conservation Service (now the Natural Resources Conservation Service, NRCS) at the US Department of Agriculture from 1993 to 1997. He served as the director of the Iowa Department of Natural Resources from 1999 to 2000. Dr. Johnson received a BS and an MS in forestry from the University of Michigan and conducted doctoral research in tropical-forest ecology in Costa Rica. He holds an honorary doctorate from Luther College in Decorah, IA. He served as a Peace Corps volunteer in Ghana from 1962 to 1964. Dr. Johnson served two terms on the National Research Council Board on Agriculture (1988–1993), where he reviewed the National Research Council report on alternative agriculture and took part in the development of the National Research Initiative Competitive Grants Program. He served as an ex officio member of the Committee on Long Range Soil and Water Conservation Policy in 1990–1993 and helped to implement many of its recommendations while chief of NRCS.

**Mark Lipson, Organic Farming Research Center, Santa Cruz, CA**

Mr. Lipson is policy program director for the Organic Farming Research Foundation. Since 1983, he has been a partner of Molino Creek Farm, a 25-acre diversified wholesale organic-vegetable operation. He served as assistant executive director for California Certified Organic Farmers from 1985 to 1992. Mr. Lipson serves on the US Department of Agriculture Advisory Committee on Agricultural Biotechnology and the Public Advisory Committee for the University of California Sustainable Agriculture Research Education Program. He has participated in a discussion panel for the National Academy of Sciences workshop on ecologic monitoring of genetically modified crops. Mr. Lipson's 1997 publication, *Searching for the O-word*, analyzes the USDA Current Research Information System for pertinence to organic farming. Mr. Lipson holds a BA in environmental studies from the University of California, Santa Cruz.

### John Miranowski, Iowa State University, Ames, IA

Dr. Miranowski is a professor in the Department of Economics at Iowa State University. He served as chair of the department in 1995–2000. Dr. Miranowski has expertise in soil conservation, natural-resource management, water quality, land management, energy, global change, and agricultural research decision-making. He has previously served as director of the Resources and Technology Division of the USDA Economic Research Service (1984–1994); executive coordinator of the secretary of agriculture's Policy Coordination Council and special assistant to the deputy secretary of agriculture (1990–1991); and Gilbert F. White fellow at Resources for the Future (1981–1982). Dr. Miranowski headed the US delegation to the Organization for Economic Cooperation and Development Joint Working Party on Agriculture and the Environment (1993–1995). He has served as a member of the Ad Hoc Working Group on Risk Assessment of Federal Coordinating Committee on Science, Education, and Technology (1990–1992); director of the Executive Board of the Association of Environmental and Resource Economists (1989–1992); and director of the Executive Board of the American Agricultural Economics Association (1987–1990). Dr. Miranowski served as a member of the National Research Council Committee on Impact of Emerging Agricultural Trends on Fish and Wildlife Habitat. He received a BS in agricultural business from Iowa State University and an MA and PhD in economics from Harvard University.

### James Moseley, US Department of Agriculture, Washington, DC

James Moseley is owner and managing partner of Infinity Pork and AgRidge Farms in Clarks Hill, Indiana. During his 32 years in farming, he has been involved in numerous public-service activities. He served as the agricultural advisor to Administrator William Reilly of the US Environmental Protection Agency from 1989 to 1990. He was assistant secretary of agriculture from 1990 to 1992. As head of the US Department of Agriculture's Soil Conservation Service and Forest Service, he was lead negotiator on issues involving endangered species, including the highly controversial spotted owl; wetlands; livestock grazing on public lands; and policy issues related to the conservation title of the 1990 farm bill. Mr. Moseley returned to farming in 1992 and was director of agricultural services and regulations for the state of Indiana at Purdue University from 1993 to 1995. Mr. Moseley served on the National Research Council Board on Agriculture and Natural Resources from 1992 to 1995. His farm operation includes 2,800 acres of no-till corn and soybeans and 50,000 hogs. Mr. Moseley's management portfolio includes a waste-treatment plant for separating and composting waste solids, chisel plowing, and construction of wildlife habitat through collaboration with Pheasants Forever. He is active in an initiative called Food, Land and People, an educational program about resources and the environment.

Mr. Moseley holds a BS in horticulture from Purdue University (1973). He resigned from the committee in April 2001 to become deputy secretary of the US Department of Agriculture.

### Elizabeth Owens, Monsanto Company, St. Louis, MO

Dr. Owens is North American co-lead for scientific affairs at the Monsanto Company. Previously, she was team lead and manager of regulatory affairs in biotechnology for potato and specialty crops. She is a recognized expert in pesticide and biotechnology regulations. Dr. Owens's previous positions include manager of government affairs (1995–1998) and manager of product registrations (1991–1995) at ISK Biosciences Corporation, manager of regulatory affairs and commercial development at BioTechnica International, Inc. (1986–1991), and Program Leader and senior scientist at GTE Laboratories (1980–1986). Dr. Owens received a BS in food science and chemistry from the University of Idaho, an MS in entomology from Iowa State University, and a PhD in entomology with an integrated pest management specialization from the University of Massachusetts at Amherst.

### David Pimentel, Cornell University, Ithaca, NY*

See About the Authors.

### Lori Ann Thrupp, US Environmental Protection Agency, San Francisco, CA

Dr. Thrupp has broad expertise in sustainable agriculture, food and environmental policy, natural-resource management, sustainable enterprise and green marketing, pesticides and pest management, and agricultural biodiversity. She is currently a life scientist at the US Environmental Protection Agency, Region 9, in the Agriculture Initiative, which entails education, collaboration, and support to the agriculture industry, farmer and nonprofit organizations, and universities on agriculture-environment issues. In 1990–1999, she served as director of sustainable agriculture at the World Resources Institute, where she coordinated and administered international sustainable-agriculture projects. She has served on the faculty for the Organization of Tropical Studies in Costa Rica and was a postdoctoral fellow at the University of California, Berkeley. Dr. Thrupp holds a BA from Stanford University and was a Marshall and Fulbright scholar at Sussex University, UK, when she received her MS and PhD in development studies, focused on agricultural development.

---

* Synthesis Committee member.

## Subcommittee on Economic and Social Development in a Global Context

### Ray A. Goldberg, Harvard Business School, Boston, MA, *Chair*

Dr. Goldberg is George Moffett professor of agriculture and business, emeritus, at the Harvard Business School. His expertise is in domestic and international agribusiness management. With John H. Davis, he developed the Agribusiness Program at Harvard Business School in 1955. From 1970 to 1997, he was the George Moffett professor of agriculture and business and head of the Agribusiness Program. As emeritus professor, he has chaired the Agribusiness Senior Management Seminars at Harvard Business School and teaches a course on agribusiness and food policy at the John F. Kennedy School of Government. Dr. Goldberg's recent publications involve developing strategies for private, public, and cooperative managers as they position their firms, institutions, and government agencies in a rapidly changing global food system. He is also conducting research on the major biologic, logistic, packaging, and information revolutions that affect global agribusiness managers as they attempt to cope with the volatile restructuring of major commodity systems. Dr. Goldberg is one of the founders and the first president of the International Agribusiness Management Association and an adviser and consultant to numerous government agencies and private firms. Dr. Goldberg received an AB from Harvard, an MBA from the Harvard Graduate School of Business Administration, and a PhD in agricultural economics from the University of Minnesota.

### Julian Alston, University of California, Davis, CA

Dr. Alston is a professor of agricultural and resources economics at the University of California, Davis. His expertise is in agricultural R&D and agricultural research policy. Specifically, Dr. Alston is interested in economics of agricultural markets and policy, agricultural sciences and technology, and international agricultural economics. Before joining the University of California, he was an agricultural economist in the Department of Agriculture in Victoria, Australia. Dr. Alston was awarded a fellowship with the American Agricultural Economics Association in 2000. Other awards he has received include outstanding published research in agricultural economics from the Western Agricultural Economics Association (1991, 1995), best article in the *Review of Marketing and Agricultural Economics* from the Australian Agricultural Economics Society (1993), and outstanding *American Journal of Agricultural Economics* article (1987). He is on the Editorial Board of *Agribusiness: An International Journal* and the *Australian Journal of Agricultural and Resource Economics*. He has also been an associate editor of the *American Journal of Agricultural Economics*. Dr. Alston received a BAgrSci from the University of Melbourne, an MAgrSci in agricultural economics from La Trobe University, and a PhD in economics from North Carolina State University.

### Christine Bruhn, University of California, Davis, CA

Dr. Bruhn is consumer food marketing specialist at the University of California Cooperative Extension in Davis. She studies consumer concerns about food safety and quality, investigates consumer response to new technologies or ingredients, identifies factors that influence consumer food choice, investigates difficulties that the food industry may have in fulfilling consumer expectations, and develops appropriate educational programs. Under the auspices of the Institute of Food Technologists, Dr. Bruhn is regional food science communicator and a member of the Food Council Advisory Committee, the Biotechnology Issues Task Force, and the Science, Communication, and Government Relations Committee. She is also chair-designate of the *Dairy, Food, and Environmental Sanitation* Journal Management Committee at the International Association for Food Protection. Dr. Bruhn received a BS and an MS in home economics and a PhD in consumer behavior from the University of California, Davis.

### Lawrence Busch, Michigan State University, East Lansing, MI

Dr. Busch is University Distinguished Professor in the Department of Sociology and director of the Institute for Food and Agricultural Standards at Michigan State University, East Lansing. His expertise includes development sociology, economic development and planning, agricultural research policy, and democratic governance. From 1988 to 1989, he was research director at the French Institute of Scientific Research for Development and Cooperation in Paris. He has served on Experiment Station Committee on Organization and Policy T5, the social-science coordinating committee for the experiment stations; the scientific advisory board for the International Center for Agronomic Research, the French agronomic-aid agency; and the review committee for the Brazilian Agricultural Research Corporation (EMBRAPA), the Brazilian equivalent of the US Department of Agriculture Research, Education, and Economics mission area. Currently, Dr. Busch serves as the US Agency for International Development scientific liaison officer to International Service for National Agricultural Research. His recent publications include *From Columbus to Conagra: The Globalization of Agriculture*; *The Agricultural Scientific Enterprise, a System in Transition*; and "Inquiry of the Public Good: Citizen Participation in Agricultural Research." He received a BA in history from Hofstra University in Hempstead, NY, and an MS and PhD in development sociology from Cornell University.

## Pierre Crosson, Resources for the Future, Washington, DC

Dr. Crosson is a senior fellow and resident consultant in the Energy and Natural Resources Division of Resources for the Future, where he analyzes agriculture and related issues of sustainability and climate-change impacts. His expertise is in agricultural, environmental, and natural-resources policy and analysis. Recent publications include *An Income and Product Account Perspective on the Sustainability of US Agriculture, Concerns for Stability: Integration of Natural Resource and Environmental Issues in the Research Agendas of NARS*, "Future Supplies of Land and Water for World Agriculture," "Demand and Supply: Trends in Global Agriculture," and "Natural Resource and Environmental Accounting in U.S. Agriculture." Dr. Crosson has served on the National Research Council Committee on the Human Dimensions of Global Change (1989–1991) and the National Research Council Panel for Collaborative Research Support for the US Agency for International Development's Sustainable Agriculture and Natural Resource Management (SANREM) Program. He has also served as a member of the National Rural Studies Committee and of a task force of the Council on Agricultural Science and Technology on Preparing the US Agriculture for Global Climate Change and as secretary of the Association of Environmental and Resource Economists. Awards that Dr. Crosson has received include the Distinguished Service Award of the Association of Environmental and Resource Economists and resident of the Rockefeller Foundation Study and Conference Center (Bellagio, Italy). Dr. Crosson received a BA from the University of Texas and a PhD in economics from Columbia University.

## Brian Halweil, Worldwatch Institute, Washington, DC

Mr. Halweil is a research associate at the Worldwatch Institute. His expertise is in food and agriculture, from organic farming to biotechnology and from hunger to water scarcity. Mr. Halweil began at the institute in 1997 as the John Gardner Public Service Fellow from Stanford University. He has written editorials and articles for *WorldWatch* magazine. He has contributed to *The State of the World* and *Vital Signs*. He was a coauthor of the Environmental Alert book *Beyond Malthus: 19 Dimensions of the Population Problem* and the Worldwatch paper "Underfed and Overfed: The Global Epidemic of Malnutrition." He has traveled extensively in Mexico, Central America, and the Caribbean, learning indigenous farming techniques and promoting sustainable food production. In Mexico and Cuba, he helped to promote biointensive farming as a means to improve household food security and household income in rural areas. Mr. Halweil received a BS in earth systems and biology from Stanford and has completed research, fieldwork, and coursework at the College of Agricultural and Environmental Sciences at the University of California, Davis.

### Fred Harrison, Jr., Fort Valley State University, Fort Valley, GA

Dr. Harrison is currently administrator and director of the Cooperative Extension Program and interim dean of the College of Agriculture, Home Economics, and Allied Programs at Fort Valley State University, where he has been a member of the faculty since 1982. Earlier, he was an assistant professor of extension education in the University of Georgia's College of Agriculture and a personnel and staff-development specialist for the University of Georgia's Cooperative Extension Service. Dr. Harrison has experience in extension and educational outreach at the county, state, national, and international levels and in the development of effective extension; and he has thorough knowledge of higher education and academic enterprise in the 1862 and 1890 land-grant systems. Dr. Harrison has been a member of the National Research Council Committee on Land-Grant Colleges of Agriculture (1993–1997) and the National Association of State University and Land-Grant Colleges (NASULGC) and chairman-elect (1992–1993) and chairman (1993–1994) of the Extension Committee on Organization and Policy (ECOP-National). Dr. Harrison received a BS in agricultural education from Fort Valley State College, an MEd in agricultural education from the University of Georgia, and a PhD in agricultural education/extension from Ohio State University.

### Carol Keiser, C-BAR Cattle Company, Inc., Carlinville, IL

Ms. Keiser has since 1985 been president of C-BAR Cattle Company, Inc., where she established and currently manages operations in feeding 3,000 cattle in custom feedlots in Texas, Kansas, Nebraska, and Northern Illinois. Ms. Keiser has experience in production agriculture and agribusiness administration. She coordinates business operations of C-BAR Cattle Co. and Lovless Feedlot with locations outside Illinois and Kansas. Since 1990, Ms. Keiser has also been president of C-ARC Enterprises, Inc., where she operates and manages Central Forage Systems, a forage marketing service for custom operators, trucking companies, and dairy and beef customers. She has expertise in agribusiness administration, policy development, research analysis, and market development. From 1997 to 1999, Ms. Keiser served as a member of the Strategic Planning Task Force on US Department of Agriculture Research Facilities. She is on the boards of directors of the Council on Food and Agricultural Research (C-FAR), the Illinois Beef Association, and Agricultural Future of America; and she was a member of the Agricultural Advisory Council for the College of Agricultural, Consumer, and Environmental Sciences and the External Advisory Council for the Department of Animal Science, University of Illinois. Ms. Keiser received a BS in agriculture from the University of Illinois and a BS in education from Greenville College.

**Terry L. Roe, University of Minnesota, St. Paul, MN**

Dr. Roe is professor of applied economics and director of the Center for Political Economy and co-director of the Economic Development Center at the University of Minnesota. His research focuses on economic growth, development, and trade. Currently funded projects include effects of agricultural trade reform on developing countries and a World Bank project on water policy for long-run growth in water-scarce developing countries. His recent publications include "Growth, Lobbying and Public Goods" and "A Global Analysis of Agricultural Reform in WTO Member Countries." Dr. Roe received a PhD in agricultural economics from Purdue University, and he was a visiting fellow at Yale University's Economic Growth Center in the academic year 1984–1985.

**Laurian J. Unnevehr, University of Illinois, Urbana-Champaign, IL***

See About the Authors.

### Subcommittee on Food and Health

**Susan Harlander, Consultant, St. Paul, MN,** *Chair*

Dr. Harlander's expertise is in food-science research and development. Dr. Harlander recently formed a consulting company on food and agricultural biotechnology. She had held the position of vice president of biotechnology development and Green Giant agricultural research at the Pillsbury Company. She has also been director of dairy foods R&D at Land O' Lakes and served on the faculty in the Department of Food Science and Nutrition at the University of Minnesota. Dr. Harlander has extensive National Research Council experience, having served as a member of the Board on Agriculture and Natural Resources, the Committee on Food Chemicals Codex, the Committee on Opportunities in the Nutrition and Food Sciences, the Ford Foundation Minority Postdoctoral Review Panel on Biological Sciences, and the Panel on the Applications of Biotechnology to Traditional Fermented Foods. Dr. Harlander is a member of the US Department of Agriculture National Agricultural Research, Extension, Education, and Economics Advisory Board and the subcommittee that oversees the Agricultural Research Service's peer-review system. She is an active member of the Institute of Food Technologists (IFT) and serves on the Foundation Board. She has previously served on IFT's Executive Committee and Committee on Research and was an IFT scientific lecturer. Dr. Harlander received a BS in biology from the University of Wisconsin-Eau Claire and an MS in microbiology and a PhD in food science from the University of Minnesota.

---

* Synthesis Committee member.

## Lester M. Crawford, Jr., Center for Food and Nutrition Policy, Washington, DC

Dr. Crawford has been director of the Center for Food and Nutrition Policy at Georgetown University since 1997. He had been executive director of the Association of American Veterinary Medical Colleges and he was executive vice-president of scientific affairs for the National Food Processors Association. He has also been administrator of the US Department of Agriculture Food Safety Inspection Service (1987–1991) and director of the Food and Drug Administration Center for Veterinary Medicine (1982–1985). Dr. Crawford was a member of the National Research Council Institute for Laboratory Animal Research (1989–1992) and has served as a National Research Council reviewer. He is on the editorial board of the *Encyclopedia of Food*, and of *Microbial Drug Resistance*, and is associate editor of *Food Control*. He is also vice president of the World Association of Veterinary Hygiene, a member of the Expert Advisory Panel on Food Safety of the World Health Organization, and a member of the Committee on Scientific Freedom and Responsibility of the American Association for the Advancement of Science. Dr. Crawford has been a consultant to industry, academe, and law firms. Recent publications include *Animal Drugs and Human Health*, "Bovine Spongiform Encephalopathy," and "Emerging Issues in Food Safety." Dr. Crawford received a DVM from Auburn University, a PhD in pharmacology from the University of Georgia and a DSc (Honoris Causa) from Budapest University.

## Joan R. Davenport, Washington State University, Prosser, WA

Dr. Davenport is associate professor and soil scientist at the Irrigated Agriculture Research and Extension Center at Washington State University (WSU). She conducts research in nutrient management, soil fertility, and management of irrigated agricultural crops in central Washington, emphasizing site-specific management, improved production, and environmental quality. She is also working on developing nondestructive crop-monitoring systems for early detection of nutrient and physiologic stresses in potato and grape and has had extensive experience in soil fertility and plant mineral nutrition. Before joining the WSU faculty in 1997, Dr. Davenport was manager of agricultural research at Ocean Spray Cranberries. She is an associate editor of HortTechnology and has served as secretary, vice chair, and chair of the Production and Management Section of the Potato Association of America and of the Mineral Nutrition Working Group of the American Society of Horticultural Science. Dr. Davenport's recent publications include "Using Site Specific Approaches to Advance Potato Management in Irrigated Systems," "Cultivar Influences Cranberry Response to Surface Sanding," and "Influences of Soil Iron and Cranberry Aerobic Status on Phosphorus Availability in Cranberry." Dr. Davenport received a BS in plant science from

Rutgers University, an MS in soil management from Iowa State University, and a PhD in soil chemistry (1985) from the University of Guelph, Ontario, Canada.

**Rebecca C. Doyle, Andrews, Doyle and Associates, Gillespie, IL**

Rebecca C. Doyle is a principal consultant at Andrews, Doyle and Associates, a food-product, agriculture, and natural-resources consulting firm. She had been the first female director of agriculture in the Illinois Department of Agriculture and Governor Edgar's point person for international marketing, biotechnology, and natural-resources protection. Ms. Doyle has been a board member of the National Association of State Departments of Agriculture, the International Agri Management Association, the Farm Foundation/Bennett Roundtable, the World Affairs Council of Central Illinois, and Women Executives in State Government. She is a partner in Hickory Grove Pork Farm and corporate officer of Belpine, Inc. (a family farm corporation), participating in production agriculture since 1979. She has been named BIO's Government Executive of the Year (1997) and *Progressive Farmer Magazine*'s Agriculture Person of the Year (1998). Ms. Doyle received a BS in agriculture communications (1975) and an MS in extension education (1977) from the University of Illinois College of Agriculture. She resigned from the committee in July 2001.

**Donald N. Duvick, Affiliate Professor, Iowa State University, Ames, IA**

Dr. Duvick was employed from 1951 to 1990 by Pioneer Hi-Bred International, Inc., successively as corn breeder and geneticist, coordinator of corn breeding, director of corn breeding, director of plant breeding, vice president of research and senior vice president of research. After retirement from Pioneer Hi-Bred, he was appointed an off-campus affiliate professor in the Department of Agronomy at Iowa State University. Dr. Duvick's research contributions over the last 50 years have been his comprehensive studies of cytoplasmic male sterility in maize, his demonstrating the relative contribution of genetics to yield gains of hybrid maize, his documenting the role of genetic diversity in plant breeding, and his proving that high-yielding hybrids have low genotype x environmental interactions. Dr. Duvick has been a member of the National Research Council Committee on Pakistan-BOSTID Agricultural Research Grants (1991–1995), the National Research Council Crop Vulnerability Work Group (1998–1993), the National Research Council Scientific Council to the Plant Gene Expression Center (1985–1993), and the National Research Council Subcommittee on Plants (1988–1989). He has also been a member of the Board of Directors of Genetic Resources Communications Systems and the boards of trustees of CIMMYT (the International Maize and Wheat Improvement Center) and the International Rice Research Institute. Dr. Duvick was elected to the National Academy of Sciences in 2002. Dr. Duvick received a BS in agriculture from the University of Illinois and a PhD in botany from Washington University, St. Louis.

**Joseph J. Jen, US Department of Agriculture, Research, Education, and Economics Mission Area, Washington, DC**

Joseph J. Jen is dean of the College of Agriculture of the California Polytechnic State University (CPSU) in San Louis Obispo. His research interests include postharvest handling of fruits and vegetables, pectin chemistry, and food enzymology. Before joining CPSU in 1992, Dr. Jen was chair of the Division of Food Science and Technology of the University of Georgia. He had worked for 6 years for Campbell Soup Company. Dr. Jen has worked as a consultant to the United Nations, US food companies, and foreign governments. He is on several boards of directors and executive committees of national and state agriculture-related organizations. He is a member of the California State Board of Food and Agriculture. Dr. Jen received a BS in agricultural chemistry from National Taiwan University, an MS in food science from Washington State University, and a PhD in comparative biochemistry from the University of California, Berkeley. He also received an MBA from Southern Illinois University. Dr. Jen resigned from the committee to become undersecretary of the US Department of Agriculture's Research, Education, and Economics Mission Area.

**John B. Kaneene, Michigan State University, East Lansing, MI**

Dr. Kaneene is professor of epidemiology and director of the Population Medicine Center at the College of Veterinary Medicine at Michigan State University. His research interests include molecular epidemiology of emerging and re-emerging enteric zoonotic diseases, epidemiology of antibiotic resistance in animal and human populations, risk-assessment modeling as related to foodborne pathogens, and epidemiology and prevention of drug residues in foods. Dr. Kaneene served on the National Research Council Subcommittee on Drug Use in Food Animals. He received a BS in mathematics (1968) and a DVM from the University of Khartoum in Sudan. He received an MPH and a PhD in epidemiology from the University of Minnesota.

**Larry Kuzminski, Duxbury, MA***

See About the Authors.

---

* Synthesis Committee member.

**Arno G. Motulsky, Professor Emeritus, University of Washington, Seattle, WA**

Dr. Motulsky is professor (emeritus-active) of medicine and genome sciences at the University of Washington, Seattle. He is considered a founder of the field of pharmacogenetics. His recent research focuses on human and medical genetics, the role of polymorphisms and mutations associated with disease susceptibilities, and molecular genetics of color vision. Dr. Motulsky has held numerous editorial positions, including positions with the *American Journal of Human Genetics* and the *Proceedings of the National Academy of Sciences*. He has served on the World Health Organization Expert Advisory Panel on Human Genetics and on various National Academy of Sciences (NAS) committees, including service as chair of the Committee on Diet and Health. He is on the NAS Temporary Nominating Group for Clinical and Epidemiological Sciences. He is a member of NAS and the Institute of Medicine. Dr. Motulsky received his BS and MD from the University of Illinois.

**David L. Pelletier, Associate Professor, Cornell University, Ithaca, NY**

Dr. Pelletier is associate professor of nutrition policy in the Division of Nutritional Sciences at Cornell University. His research, teaching, and outreach focus on the interaction of science, politics, and public values in the formation of food and nutrition policy. His research interests include agricultural biotechnology, dietary supplements and local food systems in the United States, and iron fortification, malnutrition, and child mortality in developing countries. He received a BS in biology and a BA in anthropology from the University of Arizona and an MA in anthropology and a PhD in anthropology with a nutrition minor from Pennsylvania State University. Dr. Pelletier has served on a variety of national and international advisory bodies for the US Department of Agriculture, the US Agency for International Development, the World Health Organization, the World Bank, UNICEF, and the International Food Policy Research Institute.

**Jean A.T. Pennington, Research Nutritionist, National Institutes of Health, Bethesda, MD**

Dr. Pennington is a research nutritionist in the Division of Nutrition Research Coordination at the National Institutes of Health. She had served in various positions in the Center for Food Safety and Applied Nutrition of the Food and Drug Administration. Dr. Pennington has expertise in nutrition, physiology, dietary surveillance, and food-composition databases. She has served as temporary adviser to the Food and Agriculture Organization/World Health Organization Joint Expert Committee on Food Additives and as president of the Society for Nutrition Education. Dr. Pennington received her BA in Physiology and her PhD in nutrition from the University of California, Berkeley.

## Max Rothschild, Iowa State University, Ames, IA

Dr. Rothschild is a distinguished professor of animal science at Iowa State University. His research has centered on genetic control of growth, reproduction, and health in swine, and identification of genes associated with economic traits in farm animals. Awards that Dr. Rothschild has recently received include the C.F. Curtiss Distinguished Professor in Agriculture (1999), American Association for the Advancement of Science Fellow (1998), US Department of Agriculture Honor Team Award (1997), and American Society of Animal Science Animal Breeding and Genetics Award (1995). Dr. Rothschild serves on the editorial boards of the *Journal of Animal Biotechnology*, the *Journal of Animal Breeding and Genetics*, *AgBio News and Information*, and the *Journal of Agriculture Genomics*. He serves as the coordinator of the USDA Cooperative State Research, Education, and Extension Service National Swine Genome Project, which spans a number of institutions. He is also a member of the Program Advisory Board for Agricultural Biotechnology of CAB International. Dr. Rothschild received a BS in animal science from the University of California, Davis, an MS in animal science from the University of Wisconsin, Madison, and a PhD in animal breeding (with minors in genetics and statistics) from Cornell University.

## Andrew Schmitz, University of Florida, Gainesville, FL

Dr. Schmitz is Ben Hill Griffin eminent scholar and professor of food and resource economics at the University of Florida, research professor at the University of California, Berkeley, and adjunct professor of the University of Saskatchewan. His teaching and research interests include international trade, marketing, cost-benefit analysis, and antitrust economics. He has been a consultant to many organizations, including legal firms in California, Iowa, the District of Columbia, and Canada; the National Grain and Feed Association; Sunkist Growers; Technokron; The World Bank; the US Central Intelligence Agency; the US Department of Agriculture; and Agriculture and Agri Food Canada. Recent publications include *Agricultural Policy, Agribusiness, and Rent-Seeking Behaviour*. Dr. Schmitz received a BSA and an MSc from the University of Saskatchewan and an MA and a PhD from the University of Wisconsin. He also received an earned doctor of letters degree from the University of Saskatchewan.

## John W. Suttie, University of Wisconsin, Madison*

See About the Authors.

---

* Synthesis Committee member.